Advances in Delays and Dynamics

Volume 2

Series editor

Silviu-Iulian Niculescu, Laboratory of Signals and Systems, Gif-sur-Yvette, France
e-mail: Silviu.Niculescu@lss.supelec.fr

For further volumes:
http://www.springer.com/series/11914

About this Series

Delay systems are largely encountered in modeling propagation and transportation phenomena, population dynamics and representing interactions between interconnected dynamics through material, energy and communication flows. Thought as an open library on delays and dynamics, this series is devoted to publish basic and advanced textbooks, explorative research monographs as well as proceedings volumes focusing on delays from modeling to analysis, optimization, control with a particular emphasis on applications spanning biology, ecology, economy and engineering. Topics covering interactions between delays and modeling (from engineering to biology and economic sciences), control strategies (including also control structure and robustness issues), optimization and computation (including also numerical approaches and related algorithms) by creating links and bridges between fields and areas in a delay setting are particularly encouraged.

Alexandre Seuret · Hitay Özbay
Catherine Bonnet · Hugues Mounier
Editors

Low-Complexity Controllers for Time-Delay Systems

Springer

Editors
Alexandre Seuret
Laboratoire d'Analyse et d'Architecture des Systèmes
CNRS-Université de Toulouse
Toulouse
France

Catherine Bonnet
Hugues Mounier
Laboratoire des signaux et systèmes
CNRS-Supélec-Université Paris Sud
Gif-sur-Yvette
France

Hitay Özbay
Department of Electrical and Electronics Engineering
Bilkent University
Ankara
Turkey

ISSN 2197-117X ISSN 2197-1161 (electronic)
ISBN 978-3-319-05575-6 ISBN 978-3-319-05576-3 (eBook)
DOI 10.1007/978-3-319-05576-3
Springer Cham Heidelberg New York Dordrecht London

Library of Congress Control Number: 2014936097

Printed on acid-free paper

Springer is part of Springer Science+Business Media (www.springer.com)

Preface

The purpose of this volume in the newly established series Advances in Delays and Dynamics (ADD@S) is to provide a collection of recent results on the design and analysis of *Low Complexity Controllers for Time Delay Systems*.

It is widely recognized that mathematical models of many man-made or natural dynamical systems contain time delays. Common sources of delays in engineering applications and in physical sciences are transport lag, queuing, processing and propagation time delays, as well as non-negligible computational time. For example, in data communication networks, while packets travel from source to destination they experience propagation delay due to physical distance between the two ends; processing and queuing at the routers may also contribute significantly to the end-to-end delay. Therefore, a challenging aspect of control over networks (for example tele-operations) is to compensate the effect of such time delays, which may very well be time varying. Typically, complex interconnected (networked) systems contain multiple time delays between various sub-systems and agents; these delays have be taken into account in the solution of related networked control problems such as synchronization, coordination, consensus, etc. Besides lumped time delays, where the delay value is a single parameter, distributed delay models are also used in various applications, ranging from chemical and biological processes to economics.

In the literature many different techniques have been proposed for the analysis and control of time delay systems. Widely used techniques include Lyapunov-based analysis and design in the time domain, and spectral methods in the frequency domain. The reader will find examples of these techniques in this volume. However, there is one common theme in all the chapters included here, that is, contributions towards *low complexity* controller design. Since time delay systems are special cases of infinite dimensional systems, general analysis and design methods developed for infinite dimensional systems work for systems with time delays as well. One way to obtain *low order* controllers for such systems is to design a controller for the "reduced order model of the plant." Another way is to design an infinite dimensional controller for the original plant model, then to do model order reduction for the controller, keeping track of the degradation in performance and stability robustness measures. These classical indirect methods for low order controller design have been widely discussed in the literature. Some of the papers in the present volume propose *new direct design methods* for fixed structure and low order controllers.

An important point to emphasize here is that what is meant by *low complexity* controller is not necessarily low order controller. For example, Smith predictor or similar type of controllers include a copy of the plant internally in the controller, so they are technically infinite dimensional. However, the impulse response of $(1 - e^{-Ts})/s$, appearing in a Smith predictor-based control structure, is of finite duration, i.e., it is nonzero on the time interval from 0 to T. So, its sampled-data implementation is an FIR filter whose dimension is $\lceil T/T_s \rceil$, where T_s is the sampling time. Moreover, FIR filters, being cascade banks of unit delays, have very nice numerical properties from the point of reliable implementation. Therefore, such predictor-based controllers are considered as low complexity. Similarly, although they are infinite dimensional, repetitive controllers can also be considered as low complexity.

The chapters in this volume deal with several aspects of low complexity controller design for systems with time delays. The main ideas of the individual chapters included here have been presented and discussed at the First Workshop organized by the International Scientific Research Network on Delay Systems (GDRI "*DelSys*), coordinated by the CNRS (French National Center for Scientific Research). The workshop was held in November 2012 at the *Laboratory of Signals and Systems* (a common research organization between CNRS, Supelec and University Paris-Sud), Gif-sur-Yvette, France.

The book is divided into three Parts: (I) Design Techniques, (II) Numerical Methods, and (III) Applications. Descriptions of the chapters contained in these three parts are given below.

Part I: Design Techniques

Part I of the book focusses on design techniques for time delay systems. In the following chapters, careful attention has been paid to the development of control algorithms which have a low complexity.

The first chapter by *C. Fiter, L. Hetel, W. Perruquetti* and *J.-P. Richard* deals with the development of online controller using a state-dependent sampling policy. The second chapter proposed by *H. Özbay* and *A.N. Gündes* is concerned with the design of first order controllers for unstable infinite dimensional plants. The third chapter provided by *P. Zitek, J. Bušek* and *T. Vyhlídal* proposes a new antiwindup conditioning scheme against actuator saturation in internal model control with delays. The next contribution in Part I is by *L. H. V. Nguyen* and *C. Bonnet*; this chapter considers a class of fractional neutral delay systems with an infinite number of poles asymptotic to the imaginary axis, a parameterization of all stabilizing controllers is obtained from fractional PI controllers designed earlier. The chapter by *C. Moussaoui, R. Abbou* and *J.-J. Loiseau* cope with the design of controllers for a class of delayed and constrained systems. Part I of the book ends with contributions from *A. Quadrat* and *A. Quadrat*, on the controller structure for automatic visual tracking on a mobile platform; in particular, delay effects are studied and low order controllers are compared.

Part II: Numerical Methods

Part II is dedicated to useful numerical methods for the analysis and design of controllers with reduced complexity for time-delay systems. The first paper by *S. Gumussoy* and *W. Michiels* proposes a tuning method for a fixed order and structure H_∞ controller for interconnected systems with delays. In the second contribution by *A. Baños, F. Pérez Rubio, S. Tarbouriech*, and *L. Zaccarian*, reset controllers are proposed for systems with delay in the state, that achieve delay-independent stability of the closed loop. The paper by *S. Olaru, N. Stanković, G. Bitsoris* and *S.-I. Niculescu* presents reduced complexity invariant sets for linear time-delay systems. The contribution by *D. Danciu* and *V. Răsvan* introduces new results in low order controller design using the method of lines for systems of conservation laws. In the last paper, by *A. Seuret* and *F. Gouaisbaut*, an alternative inequality to the Jensen's one is proposed to obtain tractable LMI conditions with reduced conservatism on the stability conditions, allowing then a more efficient analysis of control of delay systems by output feedback.

Part III: Applications

The last part of the book exposes several low complexity controller designs for time-delays systems with specific industrial applications such as oil-well drilling systems, haptic manipulators interfaces, and temperature control in SI engine exhaust catalyst.

The first chapter by *D. Bresh-Pietri, T. Leroy* and *N. Petit* deals with the control oriented input-delay model of the distributed temperature of a SI engine exhaust catalyst. The next two chapters focus on control of drilling systems. The one by *B. Saldivar, S.Mondié* and *A. Seuret* addresses modeling and control of the stick-slip phenomenon in oil-well drill-strings. The chapter by *T. Knüppel, F. Woittennek, I. Boussaada, H. Mounier* and *S.-I. Niculescu* propose a flatness-based control approach for suppressing vibrations occurring along a rotary oil-well drilling system. Finally, the last chapter of the book by *B. Liacu, I-C. Moraărescu, S.-I. Niculescu, C. Andriot, D. Dumur, F. Colledani* and *P. Boucher* presents low complexity control architectures for haptic interfaces subject to communication delays; the chapter includes an extensive experimental comparative study.

We would like to thank the managing editors of the Springer book series *Advances in Delays and Dynamics* (ADD@S), Dr. Thomas Ditzinger and Dr. Silviu-Iulian Niculescu, for their close collaboration and careful reading of the manuscript.

Their suggestions have significantly improved the overall quality of the present volume.

November 2013

Alexandre Seuret
Hitay Özbay
Catherine Bonnet
Hugues Mounier

Contents

Acronyms

APIOBPCS	Automatic Pipeline Inventory and Order-Based Production Control System
AWBT	Anti-Windup Bumpless Transfer
BFGS	Broyden–Fletcher–Goldfarb–Shanno (Algorithm)
BHA	Bottom Hole Assembly
BIBO	Bounded Input Bounded Output (System)
BKL	Barbashin-Krasovskii–La Salle (Principle)
BVP	(Neumann) Boundary Value Problem
CLF	Control Lyapunov Function
CNN	Cellular Neural Network
DDAE	Delay Differential Algebraic Equation
dDDE	Discrete-time Delay Difference Equation
FORE	First Order Reset Element
GM	Gain Margin
GS	Globally Stable
IAE	Absolute Error Integral
IMC	Internal Model Control
LMI	Linear Matrix Inequality
LTI	Linear-Time-Invariant (system)
MIMO	Multiple Input Multiple Output (System)
MoL	Method of Lines
MPC	Model Predictive Control
NEDC	New European Driving Cycle
ODE	Ordinary Differential Equation
PD	Proportional Derivative (Controller)
PDE	Partial Differential Equation
PI	Proportional Integral (Controller)
PID	Proportional Integral Derivative (Controller)
SI	Spark Ignited
SISO	Single Input Single Output (System)

TWC	Three Way Catalyst
UGA	Uniformly Globally Attractive
UGAS	Uniformly Globally Asymptotically Stable
WIP	Work In Progress

Part I
Design Techniques

Chapter 1
State-Dependent Sampling for Online Control

Christophe Fiter, Laurentiu Hetel, Wilfrid Perruquetti and Jean-Pierre Richard

Abstract In this chapter, we present a novel self-triggered control which aims at decreasing the number of sampling instants for the state feedback control of perturbed linear time invariant systems. The approach is based on convex embeddings that allow for designing a state-dependent sampling function guaranteeing the system's exponential stability for a desired decay-rate and norm-bounded perturbations. One of the main contributions of this work is an LMI based algorithm that optimizes the choice of the Lyapunov function so as to enlarge the lower-bound of the sampling function while taking into account both the perturbations and the decay-rate. The advantages of the approach are illustrated with a numerical example from the literature.

C. Fiter (✉)
Laboratoire d'Automatique, Génie Informatique et Signal (UMR CNRS 8219),
Université Lille 1, Villeneuve-d'Ascq, France
e-mail: christophe.fiter@univ-lille1.fr

L. Hetel
Laboratoire d'Automatique, Génie Informatique et Signal (UMR CNRS 8219),
École Centrale de Lille, Villeneuve-d'Ascq, France
e-mail: laurentiu.hetel@ec-lille.fr

W. Perruquetti · J.-P. Richard
Laboratoire d'Automatique, Génie Informatique et Signal (UMR CNRS 8219)
& INRIA Non-A Project, École Centrale de Lille, Villeneuve-d'Ascq, France
e-mail: wilfrid.perruquetti@inria.fr

J.-P. Richard
e-mail: jean-pierre.richard@ec-lille.fr

A. Seuret et al. (eds.), *Low-Complexity Controllers for Time-Delay Systems*,
Advances in Delays and Dynamics 2, DOI: 10.1007/978-3-319-05576-3_1,

1.1 Introduction

In the past decade, Networked Control Systems have become an important center of attention of researchers in control theory. These systems present numerous advantages, such as reduced wiring, plug and play devices, increased agility, and ease of maintenance. However, because they are often required to share a limited amount of resources, they generate fluctuations in the sampling intervals, which may result in unstable behaviours. This brings up new theoretical challenges.

In order to take into consideration these new difficulties, several works have studied the *stability of sampled-data systems with time-varying sampling* ([3, 6, 7, 11, 17]).

More recently, another research direction has emerged, consisting in dynamically controlling the sampling instants so as to reduce the processor and/or network loads while ensuring the desired control performances. Two main approaches can be found in the literature:

- The first approach is the *event-triggered control* one ([9, 13, 16, 18]), in which intelligent sensors send information to the controller when special events occur (for example when the system's state leaves some boundary around the equilibrium point). In most works from the literature, this approach requires a dedicated hardware.
- The second approach is the *self-triggered control* one ([1, 15, 19]), which emulates event-triggered control without dedicated hardware, by computing at each sampling instant a lower-bound of the next admissible sampling interval. In general in these works, no method is given to compute the Lyapunov function so as to optimize the lower-bound of the sampling intervals (i.e. in the worst case, independently of the system's state).

Still more recently, in [3–5], the authors proposed a third dynamic sampling control approach, *state-dependent sampling*, which consists in designing the state-dependent sampling function (i.e. the self-triggered controller) offline, thanks to a mapping of the state space and linear matrix inequalities (LMIs). An advantage was that these LMIs could be used for optimizing the Lyapunov function, thus solving the issue mentioned previously regarding regular self-triggered control approaches

In the present chapter, we present a novel self-triggered control scheme in the case of LTI systems with unknown, exogenous, state-bounded perturbations. The communication links are assumed to react instantaneously (i.e. there is no communication delay). We guarantee the system's exponential stability for a desired decay-rate using Lyapunov stability conditions and convexification arguments. One of the main contributions and advantages is the optimization of the Lyapunov function, taking into account both the effects of the sampling and the perturbation. It optimizes the lower-bound of the sampling intervals for the proposed self-triggered control scheme.

The chapter is organized as follows. First, we state the problem in Sect. 1.2 and propose the main stability result in Sect. 1.3. Then, we design the self-triggered controller in Sect. 1.4. Finally, simulation results are shown in Sect. 1.5 before concluding in Sect. 1.6. All the proofs are given in the Appendix.

Notations: $\mathbb{R}_+ = \{\lambda \in \mathbb{R}, \lambda \geq 0\}$, $\mathbb{R}^* = \{\lambda \in \mathbb{R}, \lambda \neq 0\}$, $\mathbb{N}^* = \{n \in \mathbb{N}, n \neq 0\}$, and $\mathbb{R}_+^* = \mathbb{R}_+ \setminus \{0\}$. $\lambda_{\max}(M)$ denotes the largest eigenvalue of a symmetric matrix $M \in \mathbb{R}^{n \times n}$. $\mathbb{S}_n^+$ (resp. $\mathbb{S}_n^{+*}$) is the set of positive (resp. positive definite) symmetric matrices $P \succeq 0$ (resp. $P \succ 0$) in $\mathbb{R}^{n \times n}$. The symmetric elements of a symmetric matrix are denoted by $*$. $\|.\|$ is the Euclidean norm on $\mathbb{R}^n$.

1.2 Problem Statement

1.2.1 System Description

Consider the perturbed LTI system

$$\dot{x}(t) = Ax(t) + Bu(t) + Ew(t), \ \forall t \in \mathbb{R}_+, \tag{1.1}$$

where $x : \mathbb{R}_+ \to \mathbb{R}^n$, $u : \mathbb{R}_+ \to \mathbb{R}^{n_u}$, and $w : \mathbb{R}_+ \to \mathbb{R}^{n_w}$ represent respectively the system state, the control function, and the exogenous disturbance. The matrices A, B, and E are constant with appropriate dimensions.

The control is a piecewise-constant state feedback

$$u(t) = -Kx(t_k), \ \forall t \in [t_k, t_{k+1}), \ \forall k \in \mathbb{N}, \tag{1.2}$$

where the gain matrix K is fixed and such that $A - BK$ is Hurwitz.

The sampling instants t_k are defined by the self-triggered control law

$$t_{k+1} = t_k + \tau(x(t_k)), \ \forall k \in \mathbb{N}, \tag{1.3}$$

where $\tau : \mathbb{R}^n \to \mathbb{R}_+$ is a state-dependent sampling function to be designed.

The disturbance is assumed to be state-bounded:

$$\exists W \geq 0, \ \|w(t)\|^2 \leq W \|x(t_k)\|^2, \ \forall t \in [t_k, t_{k+1}), \ \forall k \in \mathbb{N}. \tag{1.4}$$

Such a perturbation can represent model uncertainties, local nonlinearities, or some kind of measurement noises.

We denote by $\mathscr{S}$ the closed-loop system $\{(1.1), (1.2), (1.3), (1.4)\}$.

1.2.2 Objectives

Our goal is to design a sampling function τ as large as possible and with a maximal lower-bound, while ensuring the exponential stability of $\mathscr{S}$ for a desired decay-rate α, also called α-stability (i.e. such that there exists a scalar γ for which all trajectories satisfy $\|x(t)\| \leq \gamma e^{-\alpha t} \|x_0\|$ for any initial condition x_0).

To this aim, we use the well known Lyapunov exponential stability property:

Proposition 1 *(See [6] for example) Consider a scalar* $\alpha > 0$ *and a quadratic Lyapunov candidate function* $V(x) = x^T Px,\ \forall x \in \mathbb{R}^n$, *with* $P \in \mathbb{S}_n^{+*}$. *If the condition*

$$\dot{V}(x) + 2\alpha V(x) \leq 0, \tag{1.5}$$

is satisfied for all trajectories of $\mathscr{S}$, *then the system is globally* α*-stable.*

Our objectives can then be formulated as:

Objective 1: Sampling map design

Given a quadratic Lyapunov function V, design a lower-bound approximation of the optimal sampling function $\tau_{\text{opt}}^V(x) = \max \tau(x)$ such that (1.5) holds.

Objective 2: Lyapunov function design

Find a quadratic Lyapunov function V such that there exists a sampling function τ satisfying (1.5) with a minimum value $\tau^* = \inf_{x\in\mathbb{R}^n} \tau(x)$ as large as possible.

1.3 Main Stability Result

In this section, we provide sufficient stability conditions that depend on the time variable $\sigma \triangleq t - t_k$ and on the sampled-state $x \triangleq x(t_k)$.

The following result is obtained by bounding the effects of the perturbations on the system's behaviour, thanks to (1.4), by using the well known inequality ([2])

$$x^T y + y^T x \leq \varepsilon^{-1} x^T x + \varepsilon y^T y, \tag{1.6}$$

which is satisfied for any vectors x and y of same dimension, and any scalar $\varepsilon > 0$.

Theorem 1 *Consider a tuning parameter* $\varepsilon > 0$, *scalars* $\alpha > 0$ *and* $W \geq 0$, *and a sampling function* $\tau : \mathbb{R}^n \to \mathbb{R}_+$. *Then, the system* $\mathscr{S}$ *is globally* α*-stable if there exist a matrix* $P \in \mathbb{S}_n^{+*}$ *and scalars* $\eta \geq 0$ *and* $\mu \geq 0$ *such that*

$$(1 + \varepsilon^{-1})E^T PE \preceq \eta I,\ Q_5 \preceq \mu I, \tag{1.7}$$

and

$$x^T \Pi(\sigma)x \leq 0,\ \forall x \in \mathbb{R}^n,\ \forall \sigma \in [0, \tau(x)], \tag{1.8}$$

with

$$\Pi(\sigma) = \Lambda(\sigma)^T Q_1 \Lambda(\sigma) + \Lambda(\sigma)^T Q_2 + Q_2^T \Lambda(\sigma) + Q_3 + Q_4(\sigma), \tag{1.9}$$

$$\Lambda(\sigma) = I + \int_0^\sigma e^{sA} ds (A - BK), \tag{1.10}$$

$$\begin{aligned}
Q_1 &= (1+2\alpha\varepsilon)(A^T P + PA) + \varepsilon A^T PA + 2(\alpha + \varepsilon + 2\varepsilon\alpha^2)P, \\
Q_2 &= -(I + \varepsilon(A + 2\alpha I)^T)PBK, \\
Q_3 &= \varepsilon K^T B^T PBK + W\eta I, \\
Q_4(\sigma) &= \sigma W \mu \lambda_{\max}(E^T E) f_A(\sigma) I, \\
Q_5 &= A^T P + PA + \varepsilon^{-1} A^T PA + (1 + 2\alpha + \varepsilon^{-1})P,
\end{aligned} \tag{1.11}$$

and

$$f_A(\sigma) = \begin{cases} \dfrac{1}{\lambda_{\max}(A+A^T)} \left(e^{\lambda_{\max}(A+A^T)\sigma} - 1\right) & \text{if } \lambda_{\max}(A+A^T) \neq 0, \\ \sigma & \text{otherwise.} \end{cases} \tag{1.12}$$

Remark 1 Since $A - BK$ is assumed to be Hurwitz (i.e. the LTI system (1.1) without perturbation is stable with the continuous state feedback $u(t) = -Kx(t)$), one can show that there always exist parameters P, η and μ, such that the conditions (1.7) are satisfied and such that $\Pi(0) \prec 0$, provided that the tuning parameter ε, the decay-rate α, and the perturbations upper-bound W are small enough (if ε, α, and W are close to 0, then $\Pi(0) \simeq (A - BK)^T P + P(A - BK)$). With such parameters, we can thus find a sampling function τ satisfying the stability conditions of Theorem 1, which is lower-bounded by a strictly positive scalar. An algorithm enlarging that lower-bound will be provided in the next section.

1.4 Self-Triggered Controller Design

The self-triggered control scheme we propose in this chapter is based on the stability conditions from Theorem 1, which involve a few LMIs (1.7) (considering that ε is a tuning parameter) as well as the more complex set of conditions: $x^T \Pi(\sigma) x \leq 0,\ \forall x \in \mathbb{R}^n,\ \forall \sigma \in [0, \tau(x)]$.

These stability conditions can not be checked "as is" in real time, since they involve an infinite number of inequalities regarding the time variable σ. Therefore, in order to obtain checkable conditions, we propose to use:

- *A discretization of the time interval*, with a step $\Delta \in \mathbb{R}_+^*$ (which can be either chosen by the user or imposed by the system);
- *Convex embeddings* around the matrix function Π over every time interval $[j\Delta, (j+1)\Delta]$ (with $j \in \mathbb{N}$), so as to guarantee the desired performances between any two successive discrete instants. Indeed, since the matrix function Π is continuous, then for any $j \in \mathbb{N}$ it is possible to design a convex polytope with a finite number of vertices $\Pi_i^{(j)}(\Delta)$ ($i \in \{0, \ldots, N\}$) such that for all $x \in \mathbb{R}^n$,

$$\left(x^T \Pi_i^{(j)}(\Delta) x \leq 0,\ \forall i \in \{0, \ldots, N\}\right) \Rightarrow \left(x^T \Pi(\sigma) x \leq 0,\ \forall \sigma \in [j\Delta, (j+1)\Delta]\right). \tag{1.13}$$

1.4.1 Convex Embedding Design Based on Taylor Polynomials

Here, we adapt the result from [10] to design a convex embedding satisfying (1.13). The approach is based on an N-order Taylor series approximation of Π.

The vertices are defined for $i \in \{0, \ldots, N\}$ and $j \in \mathbb{N}$ as

$$\Pi_i^{(j)}(\Delta) = \hat{\Pi}_i^{(j)}(\Delta) + \nu^{(j)}(\Delta)I, \tag{1.14}$$

with

$$\hat{\Pi}_i^{(j)}(\Delta) = \sum_{k=0}^{i} \Psi_k^{(j)}(\Delta)\Delta^k, \tag{1.15}$$

$$\begin{aligned}
\Psi_0^{(j)}(\Delta) &= \Gamma_1^{(j)}(\Delta)^T Q_1 \Gamma_1^{(j)}(\Delta) + \Gamma_1^{(j)}(\Delta)^T Q_2 + Q_2^T \Gamma_1^{(j)}(\Delta) + Q_3 + L_0^{(j)}(\Delta),\\
\Psi_1^{(j)}(\Delta) &= \Gamma_2^{(j)}(\Delta)^T (Q_1 \Gamma_1^{(j)}(\Delta) + Q_2) + (\Gamma_1^{(j)}(\Delta)^T Q_1^T + Q_2^T)\Gamma_2^{(j)}(\Delta) + L_1^{(j)}(\Delta),\\
\Psi_{k\geq 2}^{(j)}(\Delta) &= \Gamma_2^{(j)}(\Delta)^T \frac{(A^{k-1})^T}{k!}(Q_1 \Gamma_1^{(j)}(\Delta) + Q_2) + (\Gamma_1^{(j)}(\Delta)^T Q_1^T + Q_2^T)\frac{A^{k-1}}{k!}\Gamma_2^{(j)}(\Delta)\\
&\quad + \Gamma_2^{(j)}(\Delta)^T \left(\sum_{i=1}^{k-1} \frac{(A^{i-1})^T}{i!} Q_1 \frac{A^{k-i-1}}{(k-i)!}\right)\Gamma_2^{(j)}(\Delta) + L_k^{(j)}(\Delta),
\end{aligned} \tag{1.16}$$

and

$$\begin{aligned}
\Gamma_1^{(j)}(\Delta) &= I + N_j(\Delta)(A - BK),\ \Gamma_2^{(j)}(\Delta) = N_j'(\Delta)(A - BK),\\
N_j(\Delta) &= \int_0^{j\Delta} \mathrm{e}^{As} ds,\ N_j'(\Delta) = AN_j(\Delta) + I.
\end{aligned} \tag{1.17}$$

If $\lambda_{\max}(A + A^T) = 0$, the matrices $L_k(j)$ are defined as

$$\begin{aligned}
L_0^{(j)}(\Delta) &= W\mu\lambda_{\max}(E^T E)(j\Delta)^2 I,\\
L_1^{(j)}(\Delta) &= 2W\mu\lambda_{\max}(E^T E) j\Delta I,\\
L_2^{(j)}(\Delta) &= W\mu\lambda_{\max}(E^T E) I,\\
L_{k\geq 3}^{(j)}(\Delta) &= 0.
\end{aligned} \tag{1.18}$$

Otherwise, if $\lambda_{\max}(A + A^T) \neq 0$, they are defined as

$$\begin{aligned}
L_0^{(j)}(\Delta) &= W\mu \frac{\lambda_{\max}(E^T E)}{\lambda_{\max}(A + A^T)} j\Delta \left(\mathrm{e}^{\lambda_{\max}(A+A^T) j\Delta} - 1\right) I,\\
L_1^{(j)}(\Delta) &= W\mu \frac{\lambda_{\max}(E^T E)}{\lambda_{\max}(A + A^T)} \left(\mathrm{e}^{\lambda_{\max}(A+A^T) j\Delta}\left(1 + j\Delta\lambda_{\max}(A + A^T)\right) - 1\right) I,
\end{aligned}$$

$$L_{k\geq 2}^{(j)}(\Delta) = W\mu \frac{\lambda_{\max}(E^T E)}{\lambda_{\max}(A+A^T)} e^{\lambda_{\max}(A+A^T)j\Delta} \left(j\Delta \frac{(\lambda_{\max}(A+A^T))^k}{k!} + \frac{(\lambda_{\max}(A+A^T))^{k-1}}{(k-1)!} \right) I. \tag{1.19}$$

Finally,

$$\nu^{(j)}(\Delta) \geq \max_{\sigma \in [j\Delta,(j+1)\Delta]} \lambda_{\max} \left(\Pi(\sigma) - \sum_{k=0}^{N} \Psi_k^{(j)}(\Delta)(\sigma - j\Delta)^k \right). \tag{1.20}$$

Remark 2 The matrices $\Psi_k^{(j)}(\Delta)$ defined in (1.16) are the *coefficients of the Taylor polynomial* of Π over $[j\Delta, (j+1)\Delta]$. The constants $\nu^{(j)}(\Delta)$ defined in (1.20) represent *upper-bounds of the Taylor series approximation error* over each time intervals $[j\Delta, (j+1)\Delta]$.

Theorem 2 *Consider $\Delta \in \mathbb{R}_+^*$, $j \in \mathbb{N}$, and $x \in \mathbb{R}^n$. The polytope vertices $\Pi_i^{(j)}(\Delta)$ defined in (1.14) satisfy (1.13): if $x^T \Pi_i^{(j)}(\Delta)x \leq 0$ for all $i \in \{0, \ldots, N\}$, then $x^T \Pi(\sigma)x \leq 0$ for all $\sigma \in [j\Delta, (j+1)\Delta]$.*

1.4.2 Design of the Sampling Function τ for Given Parameters

In this subsection, we consider that the parameters $P \in \mathbb{S}_n^{+*}$, $\eta \geq 0$, $\mu \geq 0$, and $\varepsilon > 0$ satisfying (1.7) are given.

Then, using Theorem 2, it is possible to design a lower-bound estimation τ of the maximal state-dependent sampling function satisfying the stability condition (1.8) from Theorem 1 as:

$$\tau(x) = j(x)\Delta, \tag{1.21}$$

with

$$j(x) = \min \left\{ j \in \mathbb{N} \mid \exists i \in \{0, \cdots, N\},\ x^T \Pi_i^{(j)}(\Delta)x > 0 \right\}. \tag{1.22}$$

Remark 3 It is important to guarantee that $j(x) > 0$ (and thus $\tau(x) > 0$) for any state $x \in \mathbb{R}^n$. To this aim, one may simply check if the inequality $\Pi_i^{(0)}(\Delta) \preceq 0$ is satisfied for all $i \in \{0, \ldots, N\}$. If this is not the case, it may be that the discretization step time Δ is too large, or that the tuning parameter ε, the decay-rate α, or the perturbations upper-bound W are also too large (see Remark 1). In the following subsection, we will show how to design all these parameters so as to optimize the lower-bound of the sampling function.

Remark 4 Note that the matrices $\Pi_i^{(j)}(\Delta)$ can be computed offline. The number of online computations required to perform the self-triggered control scheme (i.e. to compute $j(x)$) is upper-bounded by $n(n+1)(N+1)\frac{\tau(x)}{\Delta}$ multiplications and $(n+1)(n-1)(N+1)\frac{\tau(x)}{\Delta}$ additions. The online complexity is thus $O(\frac{\tau(x)}{\Delta}Nn^2)$, which is comparable to the one obtained in the self-triggered control scheme from [15] for example. Note that choosing a larger step Δ will reduce the number of computations, but will end in a loss of precision for the sampling function (1.21). Therefore, a trade-off has to be made between precision and online complexity.

1.4.3 Optimization of the Parameters (Maximization of the Lower-Bound τ^* of the Sampling Function)

In this subsection, we propose an LMI-based algorithm that computes the different parameters involved in the self-triggered control scheme (the Lyapunov matrix $P \in \mathbb{S}_n^{+*}$ and the additional parameters $\varepsilon > 0$, $\eta \geq 0$, and $\mu \geq 0$) so as to optimize the lower-bound of the sampling function (1.21). It is based on the following property:

Theorem 3 *Consider a tuning parameter $\varepsilon > 0$, a time-step $\Delta > 0$, and scalars $\alpha > 0$ and $W \geq 0$. Consider a matrix $P \in \mathbb{S}_n^{+*}$, scalars $\eta \geq 0$ and $\mu \geq 0$, and a sampling function $\tau : \mathbb{R}^n \to \mathbb{R}_+$ such that the conditions from Theorem 1 hold. If there exists an integer $j^* \in \mathbb{N}^*$ such that the matrix inequalities $\Pi_i^{(j)}(\Delta) \preceq 0$ are satisfied for all $i \in \{0, \dots, N\}$ and all $j \in \{0, \dots, j^* - 1\}$, then the system $\mathcal{S}$ is globally α-stable with the sampling function $\bar{\tau} : \mathbb{R}^n \to \mathbb{R}_+$ defined as $\bar{\tau}(x) = \max(\tau(x), j^*\Delta)$.*

The conditions $\Pi_i^{(j)}(\Delta) \preceq 0$ in Theorem 3 depend from the parameters P, η, μ, and ε. Therefore, it becomes clear that maximizing the sampling function lower-bound reduces to searching the parameters that maximize $\tau^* \triangleq j^*\Delta$. This can be done as follows.

Algorithm

Step 1: In this step, we consider the polytopic description (1.14) with upper-bounds on the estimation errors $\nu^{(j)}(\Delta) = 0$. Then, for a given $\varepsilon > 0$, the search for P, η and μ maximizing τ^* in Theorem 3 reduces to an LMI problem (from (1.11) and (1.14) to (1.19), we can see that the matrices $\Pi_i^{(j)}(\Delta)$ are linear in P, η and μ), which can be solved using LMI solvers. It is possible to find the appropriate parameter ε with a linear search algorithm.

Step 2: Next, we compute the value of the upper-bounds $\nu^{(j)}(\Delta)$ and of the matrices $\Pi_i^{(j)}(\Delta)$ that correspond to the obtained parameters P, η, μ, and ε.

Step 3: Finally, the lower-bound of the sampling function can be designed as

$$\tau^* = j^*\Delta,$$

with

$$j^* = \min \left\{ j \in \mathbb{N} \mid \exists i \in \{0, \cdots, N\},\ \Pi_i^{(j)}(\Delta) \succ 0 \right\}.$$

This parameter optimization presents several advantages, among which the guarantee of a large lower-bound $\tau^* = j^* \Delta$ for the sampling intervals, and the reduction of the online complexity, since it not necessary anymore to check the inequalities $x^T \Pi_i^{(j)}(\Delta) x > 0$ for $j \in \{0, \ldots, j^* - 1\}$ when computing $j(x)$ in (1.22).

1.5 Numerical Example

Consider the Batch Reactor system from [14] with some additional perturbation:

$$\dot{x}(t) = \begin{bmatrix} 1.38 & -0.20 & 6.71 & -5.67 \\ -0.58 & -4.29 & 0 & 0.67 \\ 1.06 & 4.27 & -6.65 & 5.89 \\ 0.04 & 4.27 & 1.34 & -2.10 \end{bmatrix} x(t) + \begin{bmatrix} 0 & 0 \\ 5.67 & 0 \\ 1.13 & -3.14 \\ 1.13 & 0 \end{bmatrix} u(t) + w(t),$$

$$u(t) = -\begin{bmatrix} -0.1006 & 0.2469 & 0.0952 & 0.2447 \\ -1.4099 & 0.1966 & -0.0139 & -0.0823 \end{bmatrix} x(t_k).$$

We will use the polytopic description presented in Sect. 1.4.1, with a polynomial approximation degree $N = 5$ in all the results that are presented.

1.5.1 Simulation Results

Here, we consider a time-step $\Delta = 0.005$ s, a desired decay-rate $\alpha = 0.1$, and an upper-bound $W = 0.01$ on the perturbations (i.e. such that $\|w(t)\| \leq 0.1\|x(t_k)\|$).

First, we apply the algorithm proposed in Sect. 1.4.3 to enlarge the lower-bound τ^* of the sampling function. We obtain $\tau^* = 3\Delta = 0.015$ s, with the parameters

$$P = 10^{-3} \begin{bmatrix} 0.3608 & 0.0580 & 0.2611 & -0.1701 \\ 0.0580 & 0.2443 & 0.1029 & 0.1101 \\ 0.2611 & 0.1029 & 0.2854 & -0.1314 \\ -0.1701 & 0.1101 & -0.1314 & 0.2861 \end{bmatrix}, \ \varepsilon = 0.0285,\ \eta = 0.0254, \text{ and}$$

$\mu = 0.3614$.

Then, we design the sampling function (which defines the self-triggered controller) as proposed in Sect. 1.4.2.

Figure 1.1 presents the simulation results obtained with the designed self-triggered controller and a perturbation satisfying $\|w(t)\| \leq 0.1\|x(t_k)\|$.

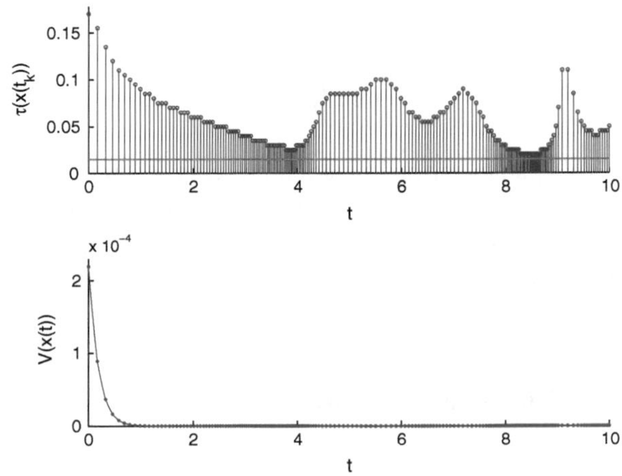

Fig. 1.1 Inter-execution times $\tau(x(t_k))$ and Lyapunov function $V(x) = x^T Px$ evolution

During this 10 s simulation, the average sampling interval is $\tau_{\text{average}} = 0.051$ s, which is more than three times the value of the optimized lower-bound τ^*.

1.5.2 Advantages of the Sampling Function's Lower-Bound Optimization

In order to illustrate the interest of the sampling function's lower-bound optimization, we consider the simpler case of the batch reactor system without perturbation nor desired decay-rate, with the same time-step $\Delta = 0.005$ s.

Using the optimization algorithm from Sect. 1.4.3, we get a lower-bound of the sampling function $\tau^* = 36\Delta = 0.18$ s, whereas with a regular approach to compute the Lyapunov function without taking into account the sampling (for example by solving the LMI $(A-BK)^T P + P(A-BK) \prec 0$, as in [14]), we get $\tau^* = 3\Delta = 0.015$ s.

Optimizing the lower-bound of the sampling function provides three main advantages.

(1) It guarantees a large lower-bound for the sampling intervals (0.18 s instead of 0.015 s in this case).

(2) It allows for using larger time-steps if one needs to reduce the number of online computations (in the present case, we can fix Δ up to 0.18 s with the optimization, and up to 0.015 s without).

(3) It allows for reducing the number of online computations, since it is not necessary to check the condition $x^T \Pi_i^{(j)}(\Delta)x \leq 0$ for $j \in \{0, \ldots, j^* - 1\}$ (here, it allows for saving $33n(n+1)(N+1)$ additions and $33(n+1)(n-1)(N+1)$ multiplications during each sampling interval... provided that the sampling interval in the non-optimized case is not smaller than $36\Delta = 0.18$ s).

1.6 Conclusion

We have introduced a novel self-triggered control approach based on convex embeddings that ensures the exponential stability with a desired decay-rate for perturbed LTI systems with linear state feedback. An interesting feature of the approach is that it allows for enlarging the lower-bound of the sampling function, by computing an appropriate Lyapunov function, thanks to LMIs. Extensions to nonlinear systems are currently under study.

Acknowledgments This work was supported by Ministry of Higher Education and Research, Nord-Pas de Calais, and FEDER through the 'CPER 2007-2013', by the European Community's Seventh Framework Programme (FP7/2007-2013) under grant agreement n° 257462: HYCON2 Network of Excellence "Highly-Complex and Networked Control Systems", and by the INTERREG IV A 2 Mers Seas Zeen Cross-border Cooperation Program 2007–2013 under SYSIASS 6–20 project.

Appendix

Proof (**Theorem 1**) Denoting $x(t) = \Lambda(\sigma)x + J_w(\sigma)$ (the solution of $\mathcal{S}$ with initial value x and perturbation w), with x the sampled-state $x(t_k)$, σ the time $t - t_k$, and the matrices

$$\Lambda(\sigma) = I + \int_0^{\sigma} \mathrm{e}^{sA} ds (A - BK), \tag{1.23}$$

and

$$J_w(\sigma) = \int_0^{\sigma} \mathrm{e}^{A(\sigma - s)} E w(s) ds, \tag{1.24}$$

one can rewrite the condition (1.5) in Proposition 1 as

$$\begin{bmatrix} \Lambda(\sigma)x + J_w(\sigma) \\ x \\ w(\sigma) \end{bmatrix}^T \begin{bmatrix} A^T P + PA + 2\alpha P & -PBK & PE \\ * & 0 & 0 \\ * & * & 0 \end{bmatrix} \begin{bmatrix} \Lambda(\sigma)x + J_w(\sigma) \\ x \\ w(\sigma) \end{bmatrix} \le 0. \tag{1.25}$$

Then, the idea of the proof is to find an upper-bound of the left part (1.25) that is independent of the perturbation w. Using the notations

$$M_1 = A^T P + PA + 2\alpha P, \; M_2(\sigma) = -PBK + M_1 \Lambda(\sigma),$$
$$M_3 = PE, \; M_4(\sigma) = E^T P^T \Lambda(\sigma),$$

we can rewrite the left part of Eq. (1.25) as

$$G_w(\sigma, x) = x^T (\Lambda(\sigma)^T M_1 \Lambda(\sigma) - \Lambda(\sigma)^T PBK - K^T B^T P \Lambda(\sigma))x$$
$$+ J_w(\sigma)^T M_1 J_w(\sigma) + J_w(\sigma)^T M_2(\sigma) x + x^T M_2(\sigma)^T J_w(\sigma)$$

$$+ J_w(\sigma)^T M_3 w(\sigma) + w(\sigma)^T M_3^T J_w(\sigma)$$
$$+ w(\sigma)^T M_4(\sigma)x + x^T M_4(\sigma)^T w(\sigma).$$

Then, using inequality (1.6), we get

$$\begin{aligned}
J_w(\sigma)^T M_2(\sigma)x + x^T M_2(\sigma)^T J_w(\sigma) &\leq J_w(\sigma)^T \varepsilon^{-1}[P + A^T PA] J_w(\sigma) \\
&\quad + x^T \varepsilon[\Lambda(\sigma)^T P \Lambda(\sigma) + [(A + 2\alpha I)\Lambda(\sigma) - BK]^T P[(A + 2\alpha I)\Lambda(\sigma) - BK]]x, \\
J_w(\sigma)^T M_3 w(\sigma) + w(\sigma)^T M_3^T J_w(\sigma) &\leq J_w(\sigma)^T P J_w(\sigma) + w(\sigma)^T E^T P E w(\sigma), \\
w(\sigma)^T M_4(\sigma)x + x^T M_4(\sigma)^T w(\sigma) &\leq w(\sigma)^T \varepsilon^{-1} E^T P E w(\sigma) + x^T \Lambda(\sigma)^T \varepsilon P \Lambda(\sigma)x,
\end{aligned}$$

and thus obtain (with matrices Q_i defined in (1.11)):

$$\begin{aligned}
G_w(\sigma, x) \leq x^T \Big[\Lambda(\sigma)^T Q_1 \Lambda(\sigma) + \Lambda(\sigma)^T Q_2 + Q_2^T \Lambda(\sigma) + \varepsilon K^T B^T P B K\Big] x \\
+ w(\sigma)^T (1 + \varepsilon^{-1}) E^T P E w(\sigma) + J_w(\sigma)^T Q_5 J_w(\sigma). \qquad (1.26)
\end{aligned}$$

Using (1.7) and (1.4), we get

$$w(\sigma)^T (1 + \varepsilon^{-1}) E^T P E w(\sigma) \leq \eta w(\sigma)^T w(\sigma) \leq W \eta x^T x. \qquad (1.27)$$

From (1.7), we also have

$$J_w(\sigma)^T Q_5 J_w(\sigma) \leq \mu \left(\int_0^\sigma \mathrm{e}^{A(\sigma - s)} E w(s) ds\right)^T \left(\int_0^\sigma \mathrm{e}^{A(\sigma - s)} E w(s) ds\right).$$

Then, using Jensen's inequality ([8], Proposition B.8), the inequality (2.2) in [12], along with classic inequalities and assumption (1.4), one gets

$$\begin{aligned}
J_w(\sigma)^T Q_5 J_w(\sigma) &\leq \sigma \mu \int_0^\sigma w(s)^T E^T \left(\mathrm{e}^{A(\sigma - s)}\right)^T \left(\mathrm{e}^{A(\sigma - s)}\right) E w(s) ds \\
&\leq \sigma \mu \int_0^\sigma \mathrm{e}^{(\sigma - s)\lambda_{\max}(A + A^T)} w(s)^T E^T E w(s) ds \\
&\leq \sigma \mu \lambda_{\max}(E^T E) \int_0^\sigma \mathrm{e}^{(\sigma - s)\lambda_{\max}(A + A^T)} \|w(s)\|^2 ds \\
&\leq \sigma W \mu \lambda_{\max}(E^T E) \left(\int_0^\sigma \mathrm{e}^{\lambda_{\max}(A + A^T)s} ds\right) \|x\|^2 \\
&= \sigma W \mu \lambda_{\max}(E^T E) f_A(\sigma) x^T x, \qquad (1.28)
\end{aligned}$$

with $f_A(\sigma)$ defined in (1.12). Implementing inequalities (1.27) and (1.28) in (1.26) shows that $G_w(\sigma, x) \leq x^T \Pi(\sigma)x$, with $\Pi(\sigma)$ defined in (1.9), and thus that Proposition 1 stability conditions are satisfied if $x^T \Pi(\sigma)x \leq 0$ for all $x \in \mathbb{R}^n$ and $\sigma \in [0, \tau(x)]$. □

Proof (**Theorem 2**) Let $\Delta \in \mathbb{R}_+^*, j \in \mathbb{N}$, and $x \in \mathbb{R}^n$.

Step (1): Compute the Taylor expansion of the matrix function Π over the time interval $[j\Delta, (j+1)\Delta]$. Remember that $\Pi(\sigma)$ involves the term $\Lambda(\sigma)$ defined in (1.10). Using the property

$$\int_0^{a+b} e^{As}ds = \int_0^a e^{As}ds + \int_0^b e^{As}ds \left(A \int_0^a e^{As}ds + I\right),$$

which is valid for any scalars a and b, we can rewrite $\Lambda(\sigma)$ as a function of $\sigma' = \sigma - j\Delta \in [0, \Delta]$:

$$\begin{aligned}\Lambda(\sigma) &= I + \left(N_j(\Delta) + \int_0^{\sigma'} e^{As}dsN_j'(\Delta)\right)(A - BK) \\ &= \Gamma_1^{(j)}(\Delta) + \sum_{i=1}^{\infty} \frac{A^{i-1}}{i!}\sigma'^i \Gamma_2^{(j)}(\Delta),\end{aligned}$$

with the notations from (1.17). Using this last equation, one can show that (the computational details are omitted)

$$\Pi(\sigma) = \sum_{k=0}^{\infty} \Psi_k^{(j)}(\Delta)\sigma'^k,$$

with the matrices $\Psi_k^{(j)}(\Delta)$ defined in (1.16). Here, the matrices $L_k^{(j)}(\Delta)$ that appear in (1.16) come from the Taylor expansion of the term $Q_4(\sigma) = \sigma W \mu \lambda_{\max}(E^T E) f_A(\sigma)I$, and are defined in (1.18) and (1.19).

A polynomial approximation of order N of Π on the interval $[j\Delta, (j+1)\Delta]$ can therefore be expressed as

$$\tilde{\Pi}_N^{(\Delta,j)}(\sigma) = \sum_{k=0}^{N} \Psi_k^{(j)}(\Delta)\sigma'^k.$$

Step (2): The approximation error term $R_N^{(\Delta,j)}(\sigma) = \Pi(\sigma) - \tilde{\Pi}_N^{(\Delta,j)}(\sigma)$ can be bounded using the relation $R_N^{(\Delta,j)}(\sigma) \preceq \nu^{(j)}(\Delta)I$, with $\nu^{(j)}(\Delta)$ defined in (1.20). With this, it is clear that if $x^T \left(\tilde{\Pi}_N^{(\Delta,j)}(\sigma) + \nu^{(j)}(\Delta)I\right) x \leq 0$, then $x^T \Delta(\sigma) x \leq 0$.

Step (3): Since the function

$$x^T \left(\tilde{\Pi}_N^{(\Delta,j)}(.) + \nu^{(j)}(\Delta)I\right) x : [j\Delta, (j+1)\Delta] \to \mathbb{R}$$

is polynomial, we can use the convex embedding design from [10], Sect. 3, to prove that if we have $x^T \Pi_i^{(j)}(\Delta) x \leq 0$ for all $i \in \{0, \cdots, N\}$, with $\Pi_i^{(j)}(\Delta) = \left(\sum_{k=0}^{i} \Psi_k^{(j)}(\Delta)\Delta^k\right) + \nu^{(j)}(\Delta)I$, then we have $x^T \left(\tilde{\Pi}_N^{(\Delta,j)}(\sigma) + \nu^{(j)}(\Delta)I\right) x \leq 0$ for all $\sigma \in [j\Delta, (j+1)\Delta]$, and therefore $\Pi(\sigma) \leq 0$. □

Proof (**Theorem 3**) The matrix inequalities and Theorem 2 ensure that the stability conditions of Theorem 1 are satisfied for the sampling function $\bar{\tau}(x) = \max(\tau(x), j^* \Delta)$. □

References

1. Anta, A., Tabuada, P.: To sample or not to sample: self-triggered control for nonlinear systems. IEEE Trans. Autom. Control **55**(9), 2030–2042 (2010)
2. Cao, Y., Sun, Y., Cheng, C.: Delay dependent robust stabilization of uncertain systems with multiple state delays. IEEE Trans. Autom. Control **43**(11), 1608–1612 (1998)
3. Fiter, C., Hetel, L., Perruquetti, W., Richard, J.-P.: A state dependent sampling for linear state feedback. Automatica **48**(8), 1860–1867 (2012)
4. Fiter, C., Hetel, L., Perruquetti, W., Richard, J.-P.: State-dependent sampling for perturbed time-delay systems. In: 51st IEEE Conference on Decision and Control, pp. 2358–2363. Maui, Hawaii, USA (2012)
5. Fiter, C., Hetel, L., Perruquetti, W., Richard, J.-P.: A robust polytopic approach for state-dependent sampling. In: 12th European Control Conference, pp. 2603–2608. Zurich, Switzerland (2013)
6. Fridman, E.: A refined input delay approach to sampled-data control. Automatica **46**(2), 421–427 (2010)
7. Fujioka, H.: Stability analysis of systems with aperiodic sample-and-hold devices. Automatica **45**(3), 771–775 (2009)
8. Gu, K., Kharitonov, V., Chen, J.: Stability of Time-Delay Systems. Birkhauser, Boston (2003)
9. Heemels, W., Donkers, M., Teel, A.R.: Periodic event-triggered control based on state feedback. In: 50th IEEE Conference on Decision and Control, pp. 2571–2576. Orlando, Florida, USA (2011)
10. Hetel, L., Daafouz, J., Iung, C.: Stabilization of arbitrary switched linear systems with unknown time-varying delays. IEEE Trans. Autom. Control **51**(10), 1668–1674 (2006)
11. Liu, K., Fridman, E.: Wirtinger's inequality and Lyapunov-based sampled-data stabilization. Automatica **48**(1), 102–108 (2012)
12. Loan, C.V.: The sensitivity of the matrix exponential. SIAM J. Numer. Anal. **14**(6), 971–981 (1977)
13. Lunze, J., Lehmann, D.: A state-feedback approach to event-based control. Automatica **46**(1), 211–215 (2010)
14. Mazo, Jr. M., Anta, A., Tabuada, P.: On self-triggered control for linear systems: guarantees and complexity. In: European Control Conference, Budapest, Hungary (2009)
15. Mazo Jr, M., Anta, A., Tabuada, P.: An ISS self-triggered implementation of linear controllers. Automatica **46**(8), 1310–1314 (2010)
16. Postoyan, R., Anta, A., Nesic, D., Tabuada, P.: A unifying Lyapunov-based framework for the event-triggered control of nonlinear systems. In: 50th IEEE Conference on Decision and Control, pp. 2559–2564. Orlando, Florida, USA (2011)
17. Seuret, A.: A novel stability analysis of linear systems under asynchronous samplings. Automatica **48**(1), 177–182 (2012)
18. Tabuada, P.: Event-triggered real-time scheduling of stabilizing control tasks. IEEE Trans. Autom. Control **52**(9), 1680–1685 (2007)
19. Wang, X., Lemmon, M.D.: Self-triggering under state-independent disturbances. IEEE Trans. Autom. Control **55**(6), 1494–1500 (2010)

Chapter 2
Design of First Order Controllers for Unstable Infinite Dimensional Plants

Hitay Özbay and A. Nazlı Gündeş

Abstract A design method for first order controllers is presented for a class of unstable infinite dimensional plants, including systems with time delays, fractional order systems, and systems represented by PDEs. The design restricts the controllers to be in the form of PI, PD and lead or lag controllers. The approach is based on the small gain theorem and requires minimization of an H_∞ norm of a transfer function over a low number of parameters. The gain margin optimization problem is solved for PD controllers. For PI controllers, optimization of the integral action gain is also discussed.

2.1 Introduction

This work deals with the design of different types of first order controllers for infinite dimensional plants whose transfer functions contain single unstable pole. In this context PI, PD, lead and lag controllers are investigated. The basic idea is to put the characteristic equation of the feedback system into a form where the small gain theorem can be applied. For this purpose, algebraic manipulations similar to those used in [6, 17] play a crucial role. Once the controller structure is fixed, the range of allowable controller gain is estimated by computing the H_∞ norm of an infinite dimensional transfer function which contains a free parameter. Optimization of this free parameter is helpful for reducing the conservative results obtained in [17].

H. Özbay (✉)
Department of Electrical and Electronics Engineering, Bilkent University,
06800 Ankara, Turkey
e-mail: hitay@bilkent.edu.tr

A. N. Gündeş
Department of Electrical Engineering, University of California, Davis, CA 95616, USA
e-mail: angundes@ucdavis.edu

A. Seuret et al. (eds.), *Low-Complexity Controllers for Time-Delay Systems*,
Advances in Delays and Dynamics 2, DOI: 10.1007/978-3-319-05576-3_2,

It should be noted that when there are only small number of free parameters in the controller, classical stability checks (e.g. Nyquist criterion) can be used to determine the set of all stabilizing controller parameters. However, this brute force method may not be numerically very attractive, especially when the plant considered is unstable and infinite dimensional.

In particular, for time delay systems there are several numerically feasible methods for finding low order controllers, see e.g. [4, 5, 14, 19]. For applications to communication networks see [11] and [20]. The method of [17] has been extended to cover fractional order systems with time delays in [16], see also [3].

This chapter is organized as follows. Section 2.2 contains several examples of engineering applications where plant model falls within the framework of the present study. A sufficient condition for the stability of the feedback system (based on the small gain theorem) is derived in Sect. 2.3. Then in Sect. 2.4 different types of controllers are designed based on this condition. Conclusions and future works are given in Sect. 2.5.

2.2 Problem Definition and Examples of Plants Considered

As mentioned above, the plants considered here have transfer functions in the form

$$P(s) = \frac{1}{s-p}\, G(s) \tag{2.1}$$

where $p \geq 0$ is the unstable pole and $G \in H_\infty$ is the stable part of the plant. Note that $G(s)$ can be irrational (plant is infinite dimensional). The factorization in the form (2.1) also implies that the plant is strictly proper.

The controllers to be designed have the following common structure

$$C(s) = K_p + \frac{K_d\, s}{\tau s + 1} + \frac{K_i}{s}\,, \qquad K_p, K_d, K_i \in \mathbb{R},\ \ \tau \geq 0. \tag{2.2}$$

Note that PD, PI, lead and lag controllers are special case of (2.2):

$$C_{pd}(s) = K_p\,(1 + \widetilde{K}_d\, s), \qquad \widetilde{K}_d = \frac{K_d}{K_p}, \tag{2.3}$$

$$C_{pi}(s) = K_p\left(1 + \frac{\widetilde{K}_i}{s}\right), \qquad \widetilde{K}_i = \frac{K_i}{K_p}, \tag{2.4}$$

$$C_\ell(s) = K_p\left(\frac{1+\alpha\,\tau\, s}{1+\tau\, s}\right), \qquad \alpha\tau = \tau + \widetilde{K}_d. \tag{2.5}$$

Clearly C_ℓ is a lead controller if $\widetilde{K}_d > 0$ and it is a lag controller if $0 > \widetilde{K}_d > -\tau$.

Definition 1 The feedback system formed by the controller C and the plant P is stable if $S := (1 + PC)^{-1}$, CS and PS are stable, i.e., they are transfer functions in H_∞. If this is the case, then the controller C is said to stabilize the plant P. The set of all controllers stabilizing a given plant P is denoted by $\mathscr{C}(P)$.

The goal of this chapter is to determine controllers $C(s)$, in the form (2.3)–(2.5), stabilizing a given unstable infinite dimensional plant $P(s)$ whose transfer function is given by (2.1). There are several applications where plant transfer functions have this structure; specific examples are given below.

Example 1 Integrating systems with transport delay:

$$P(s) = \frac{K\,e^{-hs}}{s}\,, \quad K > 0, \quad h > 0\,, \tag{2.6}$$

i.e., in this case, $p = 0$ and $G(s) = Ke^{-hs}$. There are many application examples and control methods for this plant, [12, 21]. Application examples include oil/gas pipelines, communication networks, manufacturing plants, storage systems, etc., see e.g., [13, 18].

Example 2 Abstract model of an aircraft:

$$P(s) = \frac{e^{-hs}}{s - p}\,, \quad h > 0, \quad p > 0, \quad G(s) = e^{-hs}\,. \tag{2.7}$$

This model is used for the purpose of controlling the high frequency longitudinal dynamics (short period) of an aircraft. Dynamics due to elasticity, sensor, actuator, sampling, contribute to the time delay. The product $h \cdot p$ represents how difficult it is to control this open loop unstable system. Depending on the operating regime, it is observed that $0.06 < h \cdot p < 0.37$ for an X-29 aircraft [2].

Example 3 Flexible beam with non-collocated actuator and sensor: Typically, mathematical models of flexible beams are given by partial differential equations, [1], and their transfer functions are irrational. For the free-free beam model (with normalized material parameters) shown in Fig. 2.1, the following infinite product expansion of $G(s)$ converges in H_∞ (see [9, 10]):

$$P(s) = \frac{1}{s}\,G(s)\,, \quad G(s) = \frac{2e^{-hs}}{(\tau_v s + 1)} \prod_{n=1}^{\infty} \left(\frac{1 + \varepsilon s - s^2/\omega_n^2}{1 + \varepsilon s + s^2/\widetilde{\omega}_n^2} \right)\,, \tag{2.8}$$

where $\tau_v > 0$ is the sensor parameter, $h > 0$ is the input delay, $\varepsilon > 0$ is the damping parameter of the beam and $\omega_n,\ \widetilde{\omega}_n > 0$ with $\omega_n \to 2\left(\frac{\pi}{4} + n\pi\right)^2$ and $\widetilde{\omega}_n \to \left(\frac{\pi}{2} + n\pi\right)^2$ as $n \to \infty$.

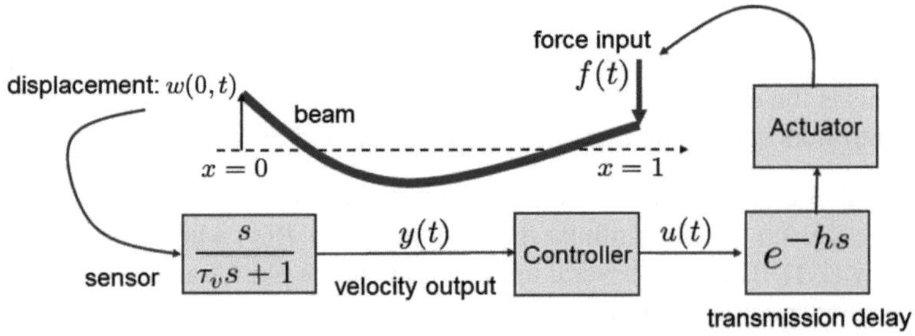

Fig. 2.1 Flexible beam control loop under delayed point force input and velocity feedback

Example 4 Interconnected systems with time delays:

$$P(s) = \frac{e^{-hs}}{s+2}\left(\frac{s+1+2(s-1)e^{-2s}}{s+1-2e^{-0.4s}}\right) = \frac{1}{s-p}G(s), \tag{2.9}$$

where $h > 0$ and $p \approx 0.5838$ is the unique root of $(s + 1 - 2e^{-0.4s}) = 0$ in $\overline{\mathbb{C}}_+$. So,

$$G(s) = e^{-hs}\left(\frac{(s+1)+2(s-1)e^{-2s}}{s+2}\right)\left(\frac{s-p}{s+1-2e^{-0.4s}}\right).$$

Example 5 A non-laminated magnetic suspension system: The following fractional order plant model is taken from [8]:

$$P(s) = \left((s^\alpha)^5 + (s^\alpha)^4 - c\right)^{-1}, \quad \alpha = 0.5, \ \ c > 0. \tag{2.10}$$

It has been shown that P can be factored as in the standard form (2.1), see [7]:

$$P(s) = \frac{1}{s-p}G(s) \ \text{ with } \ p = r^2, \ \ G(s) = \frac{(s^\alpha + r)(s^\alpha - r)}{(s^\alpha)^5 + (s^\alpha)^4 - c}$$

where $r > 0$, is the unique root of $(z^5 + z^4 - c) = 0$ on $\mathbb{R}_+$.

2.3 A Sufficient Condition for Feedback System Stability

In this section the controller is taken to be in the form C_{pd} or C_ℓ. Such a controller is stabilizing a plant in the form (2.1) if and only if there exists a constant $a > 0$ such that U_a is unimodular (i.e. $U_a, U_a^{-1} \in H_\infty$):

$$U_a(s) := \frac{s-p}{s+a} + \frac{K_p}{s+a} G(s) C_0(s),$$

where $C_0(s) := \frac{1+\alpha\tau s}{1+\tau s}$ when $C = C_{pd}$ or $C_0(s) := (1 + \widetilde{K}_d\ s)$ when $C = C_\ell$. Define

$$K_p := (p+a)G(0)^{-1} \quad \text{and} \quad G_0(s) := G(s)G(0)^{-1} \tag{2.11}$$

then

$$U_a(s) = 1 + (p+a)\frac{s}{s+a}\Psi_0(s) \quad \text{where} \quad \Psi_0(s) = \frac{1}{s}\,(\,G_0(s)C_0(s) - 1\,)\,.$$

Thus, using the fact that $\left\| \frac{s}{s+a} \right\|_\infty \le 1$, U_a is unimodular if

$$(p+a) < \|\Psi_0\|_\infty^{-1}. \tag{2.12}$$

The condition (2.12) was derived earlier in [6, 17]. Clearly, a less conservative condition for U_a to be unimodular is

$$(p+a) < \|\Psi_a\|_\infty^{-1}, \tag{2.13}$$

where

$$\Psi_a(s) := \frac{1}{s+a}\,(\,G_0(s)C_0(s) - 1\,)\,.$$

Note that

$$\|\Psi_a\|_\infty \le \|\Psi_0\|_\infty \quad \forall\, a > 0.$$

Therefore, the controller defined as above is a stabilizing controller for the plant if

$$pG(0)^{-1} < K_p < (p+a_o)G(0)^{-1},$$

where $a_o > 0$ is the largest $a > 0$ satisfying (2.13).

In order to illustrate the computations involved in the above discussion, let us consider the plant defined by (2.6) with $K = 1$ and $h > 0$. Let $C_0(s) = 1$ (i.e., consider proportional control only). The exact value of the upper bound of the controller gain can be easily computed as

$$K_p = a < \frac{\pi}{2h} \approx 1.57h^{-1}.$$

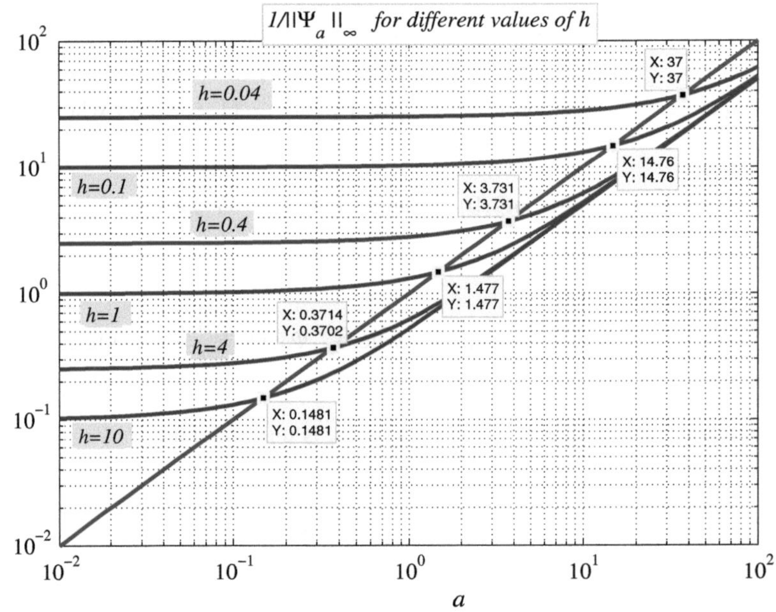

Fig. 2.2 The graph of $1/\|\Psi_a\|_\infty$ for different values of h: the largest a satisfying (2.13) is $1.48/h$

If one uses the condition (2.12), the conservative upper bound of the controller gain is

$$K_p = a < \frac{1}{\|\Psi_0\|_\infty} = h^{-1}.$$

On the other hand, if (2.13) is used, then

$$K_p = a < \tau_o h^{-1} \qquad \text{with} \qquad \tau_o \approx 1.48,$$

where τ_o is computed as shown in Fig. 2.2.

Clearly, when $C_0(s) \neq 1$, for example, $C_0(s) = \frac{1+\alpha\tau s}{1+\tau s}$ or $C_0(s) = (1+\widetilde{K}_d\, s)$, the free parameters (α, τ) or $\widetilde{K}_d$ can be used to further maximize a_o, the largest $a > 0$ satisfying (2.13).

Remark 1 There are some plants which do not admit a feasible stabilizing controller in the form C_{pd} or C_ℓ. For example, if the plant does not satisfy the parity interlacing property (PIP), then there does not exist a *stable* stabilizing controller. In order to illustrate this point, consider the plant

$$P(s) = \frac{1}{s-p}\left(\frac{1-s/z}{1+\tau s}\right) \qquad p > 0\,, z > 0, \tau > 0,$$

for which there exist a stable stabilizing controller if and only if $p < z$. If the proportional controller is defined as $C(s) = K_p = (p + a)$ then, using the notation set above,

$$\Psi_a(s) = \frac{-s\,(\tau + z^{-1})}{(s+a)(1+\tau s)} \quad \Rightarrow 1/\|\Psi_a\|_\infty = z\left(\frac{1+\tau a}{1+\tau z}\right).$$

So, the condition (2.13) becomes

$$p + a < z\left(\frac{1+\tau a}{1+\tau z}\right) = \frac{z}{1+\tau z} + \frac{\tau z}{1+\tau z}a,$$

which is stronger than the PIP, i.e., $p < z$.

2.4 PD and PI Controller Designs

2.4.1 PD Controller Design

Recall that for the plant (2.1), a PD controller is in the form $C_{pd}(s) = K_p C_0(s)$ where $K_p = (p+a)G(0)^{-1}$ and $C_0(s) = (1 + \widetilde{K}_d\, s)$. Based on the results of Sect. 2.3 the largest $a > 0$ satisfying (2.13) should be computed depending on $\widetilde{K}_d \in \mathbb{R}$. For this purpose define $G_0(s) = G(s)G(0)^{-1}$, $Q := \widetilde{K}_d \in \mathbb{R}$ and

$$\gamma(Q, a) := \left\| \frac{G_0(s) - 1}{s + a} + Q\,\frac{s}{s+a}G_0(s) \right\|_\infty . \tag{2.14}$$

In order to maximize the gain margin (GM) of the system one should try to minimize $\gamma(Q, 0)$ (the conservative approach) or try to find the largest a satisfying (2.13). See [17] for a detailed discussion on the computation of the optimal Q minimizing $\gamma(Q, 0)$ for the conservative approach. The main idea can be extended to the case $a > 0$ easily; see the algorithm given below.

Initialize: Determine a range of $Q \in [Q_{\min}\,,\, Q_{\max}] \subset \mathbb{R}$

Step 1. For each fixed Q in this interval
if it exists find the largest $a_{\max}(Q)$ such that

$$(p + a) < 1/\gamma(Q, a) \quad \forall\; a < a_{\max}(Q).$$

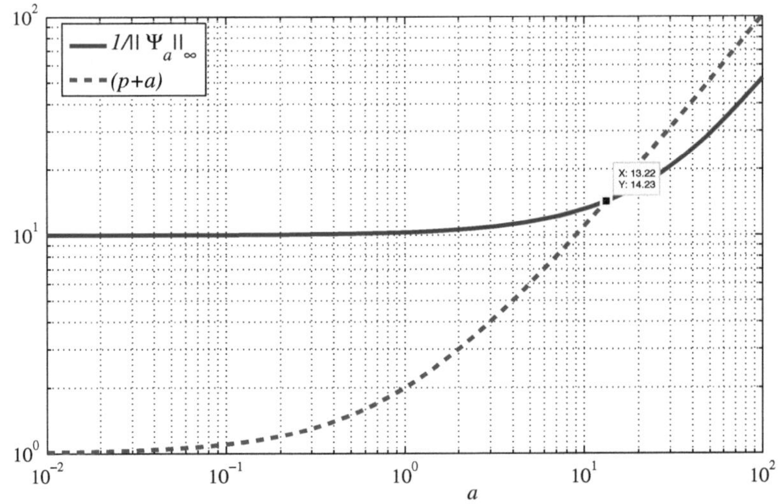

Fig. 2.3 $(p+a)$ and $1/\|\Psi_a\|_\infty$ versus a for the plant (2.7) with $h = 0.1$ and $p = 1$

Step 2. Plot Q versus $a_{\max}(Q)$ find the maximum of $a_{\max}(Q)$ and define

$$Q_{opt} := \arg\ \max\{a_{\max}(Q)\}.$$

End: An allowable range of the controller gain K_p is

$$pG(0)^{-1} < K_p < (p + a_o)G(0)^{-1} \quad \text{with} \quad a_o := a_{\max}(Q_{opt}).$$

For $p > 0$, gain margin optimizing (see [15]) PD controller parameters are

$$\widetilde{K}_{d,opt} = Q_{opt}\ , \quad K_{p,\text{GMopt}} = \sqrt{p\ (p + a_o)}G(0)^{-1}.$$

Alternatively, one can choose the *least fragile* proportional gain

$$K_{p,\text{LF}} = \left(p + \frac{a_o}{2}\right) G(0)^{-1}.$$

Step 1 of the algorithm involves drawing a graph like the one shown in Fig. 2.2. To illustrate the numerical computations, consider the plant (2.7) with $h = 0.1$ and $p = 1$. If proportional controller is used, then $Q = 0$ and $a_{\max}(0) = 13.2$ as seen in Fig. 2.3; that means the allowable range of the gain is $1 < K_p < 14.2$.

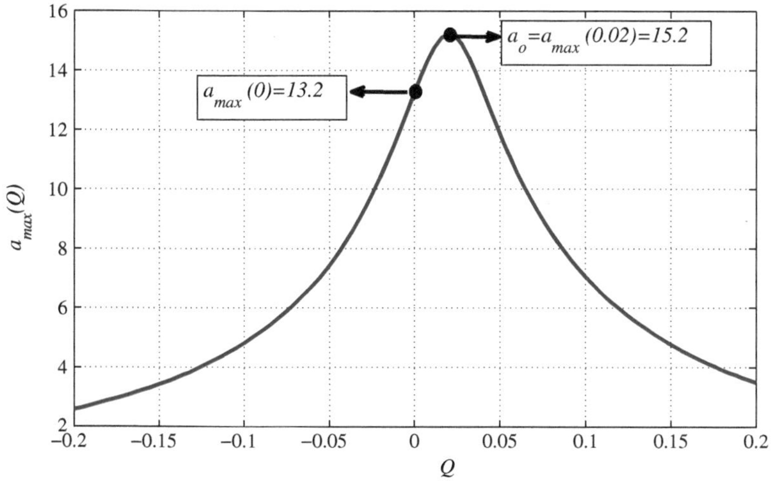

Fig. 2.4 $a_{\max}$ versus Q for the plant (2.7) with $h = 0.1$ and $p = 1$

On the other hand, it is possible to enlarge this interval by adding a derivative action. Figure 2.4 shows how $a_{\max}$ change as a function of Q. Clearly, the optimal choice is $\widetilde{K}_{d,opt} = Q_{opt} = 0.02$ and that leads to $a_o = \max a_{\max}(Q) = 15.2$ which means that the allowable gain is in the interval $1 < K_p < 16.2$ and

$$C_{pd,\mathrm{GMopt}}(s) = 4.025\,(1 + 0.02\,s) \quad \text{and} \quad C_{pd,\mathrm{LF}}(s) = 8.6\,(1 + 0.02\,s).$$

Remark 2 On Lead-Lag Controller Design.
Recall that for the lead or lag controller design $C_0(s)$ is in the form

$$C_0(s) = 1 + \frac{Q_1\,s}{1 + Q_2 s} \quad \text{with} \quad Q_1 := \widetilde{K}_d > -\tau, \quad Q_2 := \tau > 0.$$

Then, similar to the PD controller design, the parameter a which determines the controller gain should be such that $(p + a) < 1/\gamma_a$, where

$$\gamma_a(Q_1, Q_2) = \left\| \frac{G_0(s) - 1}{s + a} + \left(\frac{Q_1}{1 + Q_2\,s} \right) \frac{s}{s + a} G_0(s) \right\|_\infty .$$

So, to find $a_{\max}(Q_1, Q_2)$, in Step 1 of the corresponding gain margin optimization algorithm, the computations are done for two parameters in nested loops. Then in Step 2, a surface plot of $a_{\max}(Q_1, Q_2)$ is obtained and its maximum is determined.

2.4.2 PI Controller Design

Consider the design of a PI controller in the form

$$C_{pi}(s) = C_1(s) + \frac{K_i}{s}, \tag{2.15}$$

where $C_1(s) = K_p$ is such that $C_1 \in \mathscr{C}(P)$. In other words, a controller C_1 is already designed to stabilize P and now the integral action is to be added to the controller. The following discussion is valid for $C_1 = C_{pd}$ as well, in that case the addition of integral term will give a PID controller $C_2 = C_{pid}$.

Since $C_1 \in \mathscr{C}(P)$ the following statement holds:

$$H_1(s) := \frac{P(s)}{1 + C_1(s)P(s)} \quad \text{is in } H_\infty.$$

The characteristic equation of the feedback system formed by C_2 and P is

$$1 + C_1(s)P(s) + \frac{K_i}{s}P(s) = (1 + C_1(s)P(s))\left(1 + \frac{K_i}{s}H_1(s)\right) = 0\,.$$

Using the fact that $C_1 \in \mathscr{C}(P)$ it can be concluded that

$$C_2 \in \mathscr{C}(P) \quad \Longleftrightarrow \quad V_1^{-1} \in \mathscr{H}_\infty \;\; \text{with} \;\; V_1(s) = \left(1 + \frac{K_i}{s}H_1(s)\right).$$

Now define

$$b := K_i H_1(0)\,,$$

then V_1 can be re-written as

$$V_1(s) = \left(1 + \frac{b}{s}\right)\left(1 + \left(1 + \frac{b}{s}\right)^{-1} b\left(\frac{H_1(s)H_1(0)^{-1} - 1}{s}\right)\right). \tag{2.16}$$

Let us now assume that $b > 0$ (this is without loss of generality, since the sign of K_i can be adjusted according to the sign of $H_1(0)$). Then, note that

$$\left(1 + \frac{b}{s}\right)^{-1} = \frac{s}{s+b} \in H_\infty \quad \text{with} \quad \|\frac{s}{s+b}\|_\infty = 1.$$

The following result can be derived from the small gain theorem: $V_1^{-1} \in H_\infty$, i.e., $C_2 \in \mathscr{C}(P)$, if b satisfies

$$0 < b < 1/\|\Phi_0\|_\infty \quad \text{where} \quad \Phi_0(s) = \left(\frac{H_1(s)H_1(0)^{-1} - 1}{s}\right). \tag{2.17}$$

In fact, a careful examination of (2.16) shows that, rather than (2.17), the following less conservative sufficient condition on b can be used for C_2 to be in $\mathscr{C}(P)$:

$$0 < b < 1/\|\Phi_b\|_\infty \quad \text{where} \quad \Phi_b(s) = \left(\frac{H_1(s)H_1(0)^{-1} - 1}{s+b}\right). \tag{2.18}$$

Clearly, there is an analogy between Ψ_a and Φ_b, and the conditions (2.13) and (2.18). Note that Φ_b depends on K_p which is assumed to be in $\mathscr{C}(P)$. So, the optimal PI controller $C_{pi,opt}(s) = K_{p,opt} + (K_{i,opt}/s)$ can be designed as follows.

For each fixed $K_p \in \mathscr{C}(P)$, find the largest allowable $b > 0$ satisfying (2.18) and let it be denoted as $b_{\max}(K_p)$. Accordingly, define

$$K_{p,opt} := \arg\max\ \{\, b_{\max}(K_p) \ :\ K_p \in \mathscr{C}(P)\,\}.$$

Then, the least fragile integral action gain is

$$K_{i,opt} = \frac{b_{\max}(K_{p,opt})}{2} H_1(0)^{-1}.$$

In order to illustrate the computations involved in the design method described above, let us consider once more the plant (2.7) with $h = 0.1$ and $p = 1$. Recall from Fig. 2.3 that $C_1(s) = K_p$ is a stabilizing controller if $K_p \in (1\ ,\ 14.2)$. For each fixed K_p in this interval, define

$$H_1(s) = \frac{e^{-0.1s}}{s - 1 + K_p e^{-0.1s}}\ , \quad \text{clearly} \quad H_1(0) = \frac{1}{K_p - 1}.$$

Simple computations give Φ_b as:

$$\Phi_b(s) = \frac{H_1(s)H_1(0)^{-1} - 1}{s+b} = \left(\frac{1}{s+b}\right)\left(\frac{1 - s - e^{-0.1s}}{s - 1 + K_p e^{-0.1s}}\right).$$

Following the above procedure, for each $K_p \in (1, 14.2)$ and $b > 0$, the H_∞-norm $\|\Phi_b\|_\infty$ is computed. Then, from the graph of $1/\|\Phi_b\|_\infty$ versus b, the largest b, denoted by $b_{\max}(K_p)$, satisfying (2.18) is determined. Figure 2.5 shows $b_{\max}(K_p)$ versus K_p. Clearly, the largest $b_{\max}(K_p) = 5.7$ is achieved at $K_p = K_{p,opt} = 4.8$. For the least fragile integral gain, let $K_{i,opt} = H_1(0)^{-1} b_{\max}(K_{p,opt})/2 = 10.8$. The resulting controller is

$$C_{pi,\mathrm{LF}}(s) = 4.8 + \frac{10.8}{s}.$$

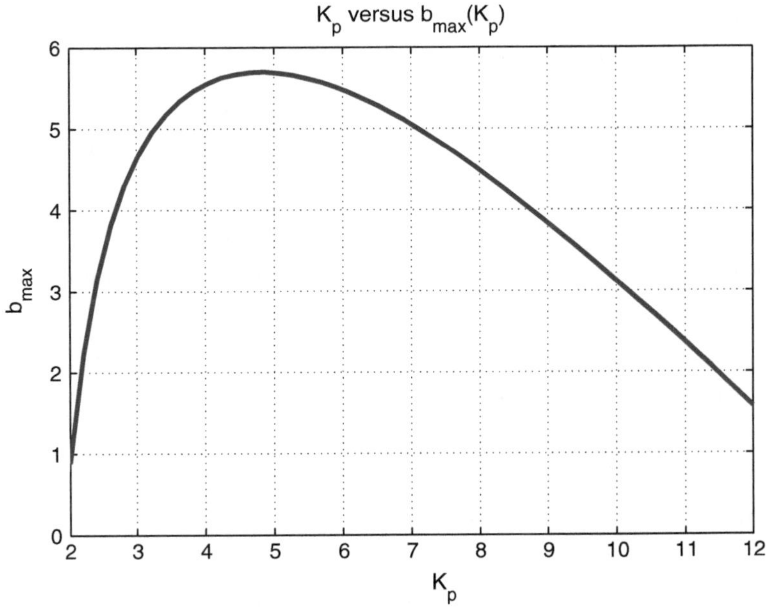

Fig. 2.5 $b_{\max}(K_p)$ versus K_p for the plant (2.7) with $h = 0.1$ and $p = 1$

2.5 Conclusions and Future Extensions

In this chapter of the book, a method is proposed for the design of stabilizing first order controllers (PD, PI and lead or lag controllers) for a class of infinite dimensional plants. The main assumption is that the plant has a single unstable pole (at the origin, or on the positive real axis). Examples from several applications are given to justify the plant model considered. These examples include systems with time delays, fractional order systems, and systems represented by PDEs.

The approach is based on the small gain theorem and requires minimization of the H_∞ norm of an infinite dimensional stable transfer function over a low number of parameters.

Another way to obtain $C_\ell(s) = K_p C_0(s)$ with a large gain margin would be to find a first order approximation of an infinite dimensional stable controller determined from the following H_∞ control problem. For a fixed $a > 0$, first, solve the one block problem

$$\gamma_o(a) = \inf_{Q \in \mathcal{H}_\infty} \| \frac{1}{s+a} \,(1 - G_0(s)Q(s))\, \|_\infty .$$

If

$$(p + a) < 1/\gamma_o(a), \tag{2.19}$$

then define

$$Q_a(s) := \arg \gamma_o(a).$$

Now, all controllers in the form $K_0 G(0)^{-1} Q_a(s)$ stabilize the plant, which is given by $P(s) = \frac{1}{s-p} G(s)$, provided that the gain is in the interval

$$p < K_0 < (p + a).$$

Thus, to maximize the allowable controller gain the maximum a defined below should be determined:

$$a_{\max} = \arg \max\{ a \ : \ a \in \mathbb{R}_+ \text{ and } (2.19) \text{ holds } \}.$$

The least fragile *stable* controller, in this framework, is

$$C_{s,LF}(s) = (p + \widehat{a}) G(0)^{-1} Q_{\widehat{a}}(s) \quad \text{where} \quad \widehat{a} := \frac{a_{\max}}{2}.$$

Approximation of $Q_{\widehat{a}}(s)$ by a first order controller, then, gives a lead or lag controller in the form $C_\ell(s)$. The above approach (and other alternative methods of approximating the plant first and then designing a low order controller) must be further compared with the proposed design of Sect. 2.4.1 on practical application examples. Currently, this is left open for a future study.

Acknowledgments This work was partially supported by the Turkish State Planning Agency (DPT) within the framework of the DPT-HAMIT project.

References

1. Curtain, R., Morris, K.: Transfer functions of distributed parameter systems: a tutorial. Automatica **45**, 1101–1116 (2009)
2. Enns, D., Özbay, H., Tannenbaum, A.: Abstract model and controller design for an unstable aircraft. AIAA J. Guidance Control Dyn. **15**, 498–508 (1992)
3. Fioravanti, A.R., Bonnet, C., Özbay, H., Niculescu, S.-I.: A numerical method for stability windows and unstable root-locus calculation for linear fractional time-delay systems. Automatica **48**, 2824–2830 (2012)
4. Gumussoy, S., Michiels, W.: Fixed-order H_∞ control for interconnected systems using delay differential algebraic equations. SIAM J. Control Opt. **49**, 2212–2238 (2011)
5. Gündeş, A.N., Özbay, H.: Low order controller design for systems with time delays. In: Proceedings of the 50th IEEE Conference on Decision and Control, Orlando, pp. 5633–5638 (2011)
6. Gündeş, A.N., Özbay, H., Özgüler, B.: PID controller synthesis for a class of unstable MIMO plants with I/O delays. Automatica **43**, 135–142 (2007)
7. Karagül, A.E., Özbay, H.: On the H_∞ controller design for a magnetic suspension system model. In: Preprints of the IFAC Joint Conference 5th SSSC, 11th TDS, 6th FDA, Grenoble, FR (2013)

8. Knospe, C., Zhu, L.: Performance limitations of non-laminated magnetic suspension systems. IEEE Trans. Control Syst. Technol. **19**, 327–336 (2011)
9. Lenz, K., Özbay, H.: Analysis and robust control techniques for an ideal flexible beam. Multidisciplinary Engineering Systems: Design and Optimization Techniques and their Applications. In: Leondes, C.T. (ed.) Control and Dynamic Systems, vol. 57, pp. 369–421. Academic Press, New York (1993)
10. Lenz, K., Özbay, H., Tannenbaum, A., Turi, J., Morton, B.: Frequency domain analysis and robust control design for an ideal flexible beam. Automatica **27**, 947–961 (1991)
11. Melchor-Aguilar, D., Niculescu, S.-I.: Computing nonfragile PI controllers for delay models of TCP/AQM networks. Int. J. Control **82**, 2249–2259 (2009)
12. Michiels, W., Niculescu, S.-I.: Stability and Stabilization of Time-Delay Systems: An Eigenvalue-Based Approach. SIAM, Philadelphia (2007)
13. Niculescu, S.-I.: Delay Effects on Stability: A Robust Control Approach. Springer, London (2001)
14. Ou, L.-L., Zhang, W.-D., Yu, L.: Low-order stabilization of LTI systems with time delay. IEEE Trans. Autom. Control **54**, 774–787 (2009)
15. Özbay, H.: Introduction to Feedback Control Theory. CRC Press LLC, Boca Raton (2000)
16. Özbay, H., Bonnet, C., Fioravanti, A.R.: PID controller design for fractional-order systems with time delays. Syst. Control Lett. **61**, 18–23 (2012)
17. Özbay, H., Gündeş, A.N.: Resilient PI and PD controller designs for a class of unstable plants with I/O delays. Appl. Comp. Math. **6**, 18–26 (2007)
18. Quet, P.-F., Ataslar, B., İftar, A., Özbay, H., Kalyanaraman, S., Kang, T.: Rate-based flow controllers for communication networks in the presence of uncertain time-varying multiple time-delays. Automatica **38**, 917–928 (2002)
19. Silva, G.J., Datta, A., Bhattacharyya, S.: PID Controllers for Time-Delay Systems. Birkhauser, Boston (2005)
20. Üstebay, D., Özbay, H., Gündeş, A.N.: A new PI and PID control design method for integrating systems with time delays: applications to AQM of TCP flows. WSEAS Trans. Syst. Control **2**, 117–124 (2007)
21. Visioli, A., Zhong, Q.: Control of Integral Processes with Dead Time. Springer, New York (2011)

Chapter 3
Anti-Windup Conditioning for Actuator Saturation in Internal Model Control with Delays

Pavel Zítek, Jaroslav Bušek and Tomáš Vyhlídal

Abstract The issue of optimizing the anti-windup scheme in the IMC design control loop with delay is dealt with in this chapter. A process model generalized on the basis of dimensional analysis is applied to obtain a broader validity of the results. The usual saturation error feedback is involved but its optimum setting is derived not from this difference but from the control error the quality of which is considered as the decisive purpose of the scheme. The optimization is performed on the basis of IAE criterion and the radial shape of this criterion lead to the final rule of setting the optimum gains in the anti-windup scheme.

3.1 Introduction

The control loop design and tuning based on linear dynamical models may exhibit a strikingly different behaviour from its implementation as soon as the always existing *actuator saturation* affects the operation. Primarily the actuating variable of the digital controller is to be artificially prevented from any possibility to exceed the saturation boundaries and particularly from any undue getting stuck at these boundaries. This faulty effect is referred to as *windup* and the schemes getting the controller saturation rid of this fault are considered as *anti-windup* schemes.

The windup problem results from the controversy between the inevitable actuator saturation and the required integrating nature of the controller action. The history

P. Zítek (✉) · J. Bušek · T. Vyhlídal
Department of Instrumentation and Control Engineering, Faculty of Mechanical Engineering, Czech Technical University in Prague, Technická 4, 166 07 Praha 6, Czech Republic
e-mail: pavel.zitek@fs.cvut.cz

J. Bušek
e-mail: jaroslav.busek@fs.cvut.cz

T. Vyhlídal
e-mail: tomas.vyhlidal@fs.cvut.cz

A. Seuret et al. (eds.), *Low-Complexity Controllers for Time-Delay Systems*, Advances in Delays and Dynamics 2, DOI: 10.1007/978-3-319-05576-3_3,

of anti-windup arrangements has gone through a variety of opinions and schemes of digital techniques in controller implementation. Already in 1967 a study on anti-windup issue in digital PID control loop was presented by Fertik and Ross [1]. Most of the proposed anti-windup schemes are based on an observer-like state feedback closed from the *saturation error* i.e. from the difference between the original and the saturated signal. These observer-like schemes have been investigated by Åström and Rundquist [2] and later by Kapoor et al. [3]. A survey of anti-windup schemes, based on the observer theory, with the so-called bumpless transfer compensation (AWBT) has been presented by Kothare et al. [4]. The observer-like feedback added to the controller serves to estimate the controller states in the face of actuator saturation and in this way, as soon as possible to restore a consistency between both the controller state and its constrained output.

A more general approach to the anti-windup issue has been introduced by Hanus et al. [5] and Doyle et al. [6] as the *conditioning technique*. The crucial aim of this technique is to get the controller state back to the normal mode as promptly as possible and to minimize the intervals of saturation effects. In [4] a unified conditioning is proposed as an achievement of "graceful deterioration" of the closed-loop performance due to implementing the actuator saturation. The conditioning schemes were further investigated and developed by Edwards and Postlethwaite [7], and Weston and Postlethwaite [8] with the extension towards the multivariable systems. A low-order observer scheme is presented in [9] and the robustness issue of the anti-windup scheme design was investigated in [10].

Both the general assumptions, namely that the control loop performance *deteriorates* due to the actuator saturation and that the non-constrained actuating signal is to be restored as promptly as possible are well-founded in the most considered cases of control loop performance. But in case of the *disturbance rejection* the impact of saturation is not always deteriorating, even an improved response due to saturation may sometimes be encountered. Also a lingering some time at the saturation boundary may also be of benefit for the control performance. As foreseen in [5] the anti-windup conditioning schemes should be viewed not only as the actuator constraint but rather as a means of the control loop tuning to some extent.

The anti-windup strategies and conditioning techniques are well elaborated for the control loops with plants and controllers considered as rational transfer functions. On the other hand, some issues remain still open in controllers involving the delay operation marked out with meromorphic transfer functions. A special kind of these controllers results from applying the scheme of Internal Model Control (IMC) to time delay systems [11]. An early investigation of the anti-windup issue in the scheme of IMC is investigated in [12]. An innovation of conditioning technique for controller operating with time delay procedure developed on the IMC principle is presented in this chapter. The rest of the chapter is structured as follows. A dimensionless generic model of the time delay plant is introduced in Sect. 3.2 and the corresponding IMC based controller given by a meromorphic transfer function is developed in Sect. 3.3. The utilizing of ultimate frequency in the controller design is dealt with in Sect. 3.4 and a novel windup-observer conditioning for this kind of controller is proposed in Sect. 3.5. Its tuning optimizing the IAE of control error is explained in Sect. 3.6. An application example and a summary to the observer tuning are given in Sect. 3.7 and some concluding remarks are added in Sect. 3.8.

3.2 Dimensionless Model of the Considered Plant

For a general investigation of the efficiency of the anti-windup schemes a sufficiently generic model of the plant is needful. Assume a stable linear time delay plant described by second order differential equation

$$\frac{d^2y(t)}{dt^2}+a_1\frac{dy(t)}{dt}+a_0y(t)=Ka_0u(t-\tau), \tag{3.1}$$

to express the dynamics of a rather wide class of stable processes free of RHP zero effects. In [13] this model is rearranged to a generic dimensionless form by means of the following dimensionless *similarity numbers*

$$\lambda=\frac{a_0}{a_1^2},\quad \vartheta=a_1\tau, \tag{3.2}$$

introduced as the so-called *swingability* and *laggardness* similarity numbers respectively. The dimensional analysis leading to this selection is presented in [13]. To obtain a totally dimensionless model also time t is to be replaced by the ratio $\bar{t}=t/\tau$. Then applying these variables and parameters in (3.1) all plants of this type are described by a common dimensionless model

$$\frac{d^2y(\bar{t})}{d\bar{t}^2}+\vartheta\frac{dy(\bar{t})}{d\bar{t}}+\vartheta^2\lambda\, y(\bar{t})=K\vartheta^2\lambda\, u(\bar{t}-1), \tag{3.3}$$

and the plants with the same λ and ϑ are referred to as *dynamically similar*. It means that for a pair of such plants whose $u(\bar{t})$ are identical their responses $y(\bar{t})$ are *identical* as well (e.g. the step responses considered in the common relative time $\bar{t}$). Another advantage of model (3.3) is the reduced number of parameters: instead of four parameters in (3.1) only three numbers λ, ϑ, K determine the set of dynamically similar plants, i.e. the plants with different a_0, a_1, τ, K but with the same λ, ϑ, K.

3.2.1 *Identification of the Plant Model with a Given Step Response*

The modification of (3.1) to the form (3.3) has been chosen as a generic pattern of the plant. As regards its application potential one has to be aware of which type of plants it may be identified with. Suppose that the plant dynamics are given by a (measured) step response $h(t)$, considered in the relative time as $h(\bar{t})$. The plant may be widely variable, either aperiodic or oscillatory and characterized by a dead time. We only assume that $h(\bar{t})$ is not affected by typical effects of the RHP zeros.

Simultaneously with the introduced time ratio $\bar{t} = t/\tau$ in (3.3) we have to adopt an adequate modification of the Laplace transform operator $s \rightarrow \bar{s} = s\tau$, and all the transfer functions to consider in this dimensionless complex variable. Now, let the usual method of successive integrals of $h(\bar{t})$ be applied in identifying adequate λ, ϑ parameters to the given step response. The following Lemma holds for this identification.

Lemma 1 *Suppose that for a stable SISO time delay process its unit-step response $h(\bar{t})$ is given as a function $h(\bar{t})$ of the dimensionless time $\bar{t} = t/\tau$. Such dimensionless parameters K, λ, ϑ of model* (3.3) *are to be assessed that $h(\infty)$ and the limit values* $\lim_{\bar{t}\to\infty} h_I(\bar{t})$ *and* $\lim_{\bar{t}\to\infty} h_{II}(\bar{t})$ *of the following integrals*

$$h_I(\bar{t}) = \int_0^{\bar{t}} [h(\infty) - h(\sigma)]\, d\sigma \quad \text{and} \quad h_{II}(\bar{t}) = \int_0^{\infty} \left[h_I(\infty) - h_I(\bar{t})\right] d\bar{t}, \tag{3.4}$$

are of the same values for $h(\bar{t})$ and for the model. The model parameters satisfying these conditions are as follows

$$K = h(\infty), \vartheta = \frac{S_1}{S_1^2 + S_1 - S_2}, \lambda = \frac{S_1^2}{S_1^2 + S_1 - S_2}, \tag{3.5}$$

where $S_1 = h_I(\infty)/h(\infty) - 1$ *and* $S_2 = h_{II}(\infty)/h(\infty) - 0.5$.

Proof The proof of this Lemma is based on investigating the step response Laplace transform of the plant (3.3)

$$\bar{H}(\bar{s}) = \frac{K \exp(-\bar{s})}{\bar{s}\left(\bar{s}^2(\lambda\vartheta^2)^{-1} + \bar{s}(\lambda\vartheta)^{-1} + 1\right)}, \tag{3.6}$$

and its integrations. The proof is presented in [13] and therefore it is omitted here. □

3.3 Internal Model Controller Design

The schemes of anti-windup and conditioning are well developed for the controllers with the rational transfer function structure [4]. However, the specific design methods for the time delay plants also lead to controllers whose transfer functions may become meromorphic, i.e. involving delayers in their structure. Our chapter is focused on the anti-windup issue just for this class of controllers. As a particular case of Youla affine parameterization the most specific method in this area is the well-known Internal Model Control (IMC) [14]. The internal plant model corresponding to (3.3) is supposed as $G_M(\bar{s})$ and is considered in a parallel linkage with the plant while the feedback is formed by the control function $Q(\bar{s})$. It holds for this parameterization that if $Q(\bar{s})$ is any proper and stable function and the plant is stable then the equivalent controller $R(s)$ of classical control loop is parameterized as follows

$$R(\bar{s}) = \frac{Q(\bar{s})}{1 - Q(\bar{s})G_M(\bar{s})}, \tag{3.7}$$

and this control loop is always internally stable [14]. On the other hand the plant stability is a necessary condition of the direct application of the IMC design.

The selection of $Q(\bar{s})$ is based on the idea of inverting the plant model and therefore an inner-outer factorization of $G_M(\bar{s})$, $G_M(\bar{s}) = G_I(\bar{s})G_O(\bar{s})$ is necessary, separating the outer factor $G_O(\bar{s})$ as invertible. Then the applicable $Q(\bar{s})$ is considered in the following manner

$$Q(\bar{s}) = [G_O(\bar{s})]^{-1} F(\bar{s}), \tag{3.8}$$

where a stable filter transfer function $F(\bar{s})$, with the unit static gain $F(0) = 1$, is selected to predetermine the desirable dynamics of the control closed loop. To predetermine a dimensionless conjugate pair $\bar{p}_{1,2} = \Phi(-\delta \pm j)$ as the dominant poles of the control loop this function is to be selected as

$$F(\bar{s}) = \frac{\Phi^2(\delta^2 + 1)}{(\bar{s} + \delta\Phi)^2 + \Phi^2}. \tag{3.9}$$

Apparently Φ in $\bar{p}_{1,2}$ is a *frequency angle* connected with the desirable natural frequency Ω of the control loop, $\Phi = \Omega\tau$, and δ is the *damping ratio* of the corresponding oscillations. Since the invertible factor of model (3.3) is $G_O(s) = K\vartheta^2\lambda(s^2 + \vartheta s + \vartheta^2\lambda)^{-1}$ the control function (3.8) results as

$$Q(\bar{s}) = \frac{(\bar{s}^2 + \vartheta\bar{s} + \vartheta^2\lambda)(\delta^2 + 1)\Phi^2}{K\vartheta^2\lambda[(\bar{s} + \delta\Phi)^2 + \Phi^2]}. \tag{3.10}$$

Then from the parameterization (3.7) the following meromorphic controller is obtained

$$R(\bar{s}) = \frac{(\bar{s}^2 + \vartheta\bar{s} + \vartheta^2\lambda)(\delta^2 + 1)\Phi^2}{K\vartheta^2\lambda\,[\bar{s}^2 + 2\delta\Phi\,\bar{s} + (1 + \delta^2)\Phi^2(1 - \exp(-\bar{s}))]}. \tag{3.11}$$

Notice that nothing but the plant parameters λ, ϑ, K and the $\bar{p}_{1,2}$ specification by Φ, δ are used in this formula. The controller (3.11) provides the well known property of compensating the control loop for the time delay in the ideal case that internal model $G_M(\bar{s})$ is just equal to the real plant transfer function. If this equivalence is achieved the tracking transfer function of the control loop is of the form

$$T(\bar{s}) = \frac{R(\bar{s})G(\bar{s})}{1 + R(\bar{s})G(\bar{s})} = \frac{(1+\delta^2)\Phi^2 \exp(-\bar{s})}{(\bar{s}+\delta\Phi)^2 + \Phi^2}. \tag{3.12}$$

In spite of the plant delay this function does not have other poles but the pair $\bar{p}_{1,2}$. Nevertheless, this desirable property ceases to hold as soon as the internal model differs from the real plant properties. Then the behaviour of the control loop is the more different from (3.12) the more different the model $G_M(\bar{s})$ from the real plant.

The controller transfer function (3.11) is not rational it involves a delay operation in controller structure and its meromorphic nature has to be considered in the design of anti-windup scheme too. For this reason the state formulation of $R(\bar{s})$ is needed. A possible conversion of controller (3.11) to an equivalent state description can be obtained by the method of *nested integrations* with the following result.

Proposition 1 *Consider the controller* (3.11) *given by the plant parameters* ϑ, λ *and the prescribed filter parameters* δ, Φ *and suppose zero initial conditions of both the models. In the L-transform the state formulation equivalent to* (3.11) *is given by the following state and output equations respectively*

$$\bar{s}x(\bar{s}) = A(\bar{s})x(\bar{s}) + B(\bar{s})e(\bar{s}), \tag{3.13}$$

$$u(\bar{s}) = Cx(\bar{s}) + De(\bar{s}), \tag{3.14}$$

where the matrices are as follows

$$A = \begin{bmatrix} 0, & -(1+\delta^2)\Phi^2(1-\exp(-\bar{s})) \\ 1, & -2\delta\Phi \end{bmatrix}, \quad B = \begin{bmatrix} 1 - \frac{(1+\delta^2)\Phi^2}{\vartheta^2\lambda}(1-\exp(-\bar{s})) \\ \frac{1}{\vartheta\lambda}\left(1 - \frac{2\delta\Phi}{\vartheta}\right) \end{bmatrix},$$

$$C = \left[0, \ \frac{(1+\delta^2)\Phi^2}{K}\right], \qquad D = \frac{(1+\delta^2)\Phi^2}{K\vartheta^2\lambda}. \tag{3.15}$$

Proof The following transform equation

$$\begin{aligned} K\vartheta^2\lambda\left[\bar{s}^2 + 2\delta\Phi\bar{s} + (1+\delta^2)\Phi^2(1-\exp(-\bar{s}))\right]u(\bar{s}) \\ = \left[(\bar{s}^2 + \vartheta\bar{s} + \vartheta^2\lambda)(\delta^2+1)\Phi^2\right]e(\bar{s}), \end{aligned} \tag{3.16}$$

is given by the controller transfer function (3.11) for the case of the zero-valued initial conditions. After cancelling the factor $\vartheta^2\lambda(1+\delta^2)\Phi^2$ in $R(\bar{s})$ the first state equation results from the terms without $\bar{s}$ in (3.11)

$$\bar{s}x_1(\bar{s}) = e(\bar{s}) - K\left[1 - \exp(-\bar{s})\right]u(\bar{s}). \tag{3.17}$$

After inserting this substitution into (3.16) the second state variable results from the transform equation

$$\bar{s}x_2(\bar{s}) = x_1(\bar{s}) + \frac{1}{\vartheta\lambda}e(\bar{s}) - \frac{2\delta K}{(1+\delta^2)\Phi}u(\bar{s}). \tag{3.18}$$

By the help of substitutions (3.17) and (3.18) the original Eq. (3.16) is reduced to the equality

$$u(\bar{t}) = \frac{(1+\delta^2)\Phi^2}{K}\left[x_2(\bar{t}) + \frac{e(\bar{t})}{\vartheta^2\lambda}\right], \tag{3.19}$$

by means of which the matrices as in (3.15) result from (3.17) and (3.18). This model holds as long as the actuating variable remains within its *operating range*, and as soon as any of the *saturation boundaries* are reached an anti-windup scheme is to be applied.

The following characteristic equation of the controller results from matrix $A(\bar{s})$ in (3.15)

$$\det\left[\bar{s}I - A(\bar{s})\right] = \bar{s}(\bar{s} + 2\delta\Phi) + (1+\delta^2)\Phi^2[1 - \exp(-\bar{s})] = 0, \tag{3.20}$$

from where it is apparent that the controller dynamics are given only by prescribing the filter poles by Φ and δ independently of the plant properties. Due to the factor $\left[1 - \exp(-\bar{s})\right]$ the value $\bar{s} = 0$ is always a solution of (3.20) being the rightmost pole of (3.11) which provides its *integrating* character. □

3.4 Utilizing the Ultimate Frequency in the IMC Design

So far frequency angle Φ was supposed to be an optional parameter of the filtering function $F(\bar{s})$ used in the IMC design. On the other hand, in (3.12) we saw that in case of a well fitting model angle Φ determines also the natural frequency $\Omega = \Phi/\tau$ of the control loop, i.e. the imaginary part of $\bar{p}_{1,2} = \Phi(-\delta \pm j)$ which are supposed to be the dominant ones. However, from the dominant pole placement techniques it is well known that just the *ultimate frequency* ω_k of the plant plays a key role in tuning the controller. Particularly with regard to reaching the poles $\bar{p}_{1,2}$ as dominant, it is advisable to suppose the ultimate frequency as the well fitting natural frequency for the control loop. Attempts to prescribe Ω higher than ω_k often result in emerging an undesirable lower frequency in the system response which spontaneously becomes dominant instead of the prescribed one.

The ultimate gain and the corresponding ultimate frequency result from the characteristic equation of plant (3.3). After adding a proportional feedback with the gain K_k to (3.3) the following characteristic equation of the loop is obtained

$$\bar{s}^2 + \vartheta\,\bar{s} + \vartheta^2\lambda + K\vartheta^2\lambda K_k \exp(-\bar{s}) = 0. \tag{3.21}$$

As it results from the dimensional analysis in [13] the *ultimate angle*, $\Phi_k = \omega_k\tau$, is the *similarity number* corresponding to the ultimate frequency. With respect to the Buckingham's theorem a dimensionless relationship has to exist between Φ_k of plant (3.3) and the parameters ϑ, λ, K.

Theorem 1 *For any pair of different but similar plants* (3.3) *with the same* $\vartheta > 0$, $\lambda > 0$, K *it holds that their ultimate angle* Φ_k *is also the same, independent of* K. *It is given by the least positive root of the equation*

$$\tan \Phi_k = \frac{\vartheta \, \Phi_k}{\Phi_k^2 - \vartheta^2 \lambda} \quad \rightarrow \quad \Phi_k = \Phi_k(\lambda, \ \vartheta). \tag{3.22}$$

Proof The proof of this relationship is provided by means of considering the undamped oscillations at frequency ω_k, i.e. $\bar{s} = j\omega_k \tau = j\Phi_k$. The proof is omitted since it can be found in [13]. □

As regards (3.22) it is important that only λ, ϑ are sufficient to determining the ultimate angle Φ_k, the influence of K is cancelled. Basically, Eq. (3.22) admits infinitely many real solutions owing to the periodicity of the tangent function. However, with respect to both the periodical character of tangent function and the physical meaning of the ultimate frequency only the smallest of the positive roots Φ_k of (3.22) can represent the ultimate angle and consequently the ultimate frequency ω_k as well.

In investigating the function $\Phi_k = \Phi_k(\lambda, \vartheta)$ it should be noted that the value of the *laggardness number* ϑ fall only into a rather narrow interval, approximately $\vartheta \in \langle 0.5, 3 \rangle$ where the upper bound of this range is to exclude *too large delays*. The values $\vartheta > 3$ would bring about a situation when a single loop controller feedback action unavoidably comes *too late* to compensate the impact of disturbances and obviously it is meaningless to consider such options. On the other hand the lower bound of ϑ prevents model (3.3) from its application to a plant with *too small delays*. Too small value of ϑ ($\vartheta < 0.5$) indicates that the delay is a marginal property of the plant and then it is suitable to replace model (3.3) rather by a delay-free model.

Also the *swingability number* λ values are to be limited. The upper bound $\lambda \leq 2$ is considered to exclude the plants with excessively weakly damped oscillations which are out of the scope of this chapter. Owing to the constrained λ and ϑ the values of ultimate angle $\Phi_k = \Phi_k(\lambda, \vartheta)$ are also from a relatively narrow interval, $\Phi_k \in \langle 0.6, 2.6 \rangle$ approximately. Notice that the value $\Phi_k = 0.5\pi$, singular for $\tan \Phi_k$ lies within this interval, but this singularity is of the so-called *removable type* and as shown in Fig. 3.1, despite this singularity the function $\Phi_k = \Phi_k(\lambda, \vartheta)$ is continuous within its whole range.

Recall now the IMC controller design (3.11) again. If the ultimate angle Φ_k is set as $\Phi = \Phi_k$, and if the above introduced relationship $\Phi_k = \Phi_k(\lambda, \vartheta)$ is applied the whole IMC control loop composed of (3.3) and (3.11) is determined by the similarity numbers λ, ϑ and the damping ratio δ only, while the gain parameter K is cancelled. If moreover δ is kept constant any of the considered control loop performance properties is unambiguously tied with a pair of λ, ϑ values. Therefore it is evident that any performance measure of the considered control loop can be

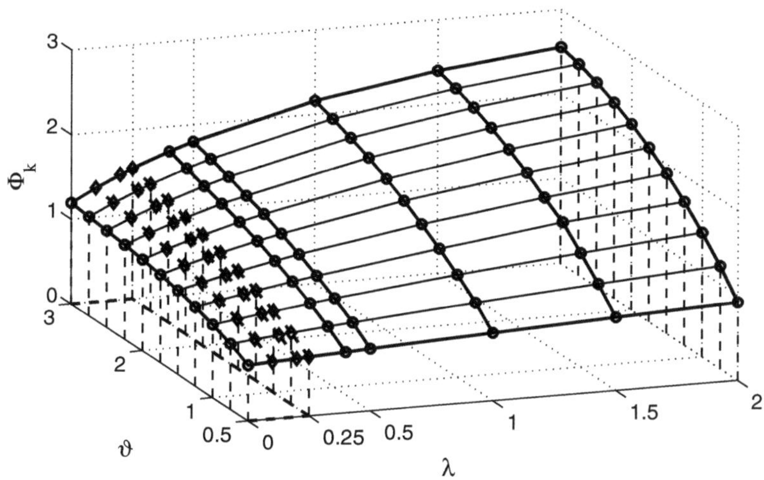

Fig. 3.1 3D plot of relationship (3.22)

mapped over the admissible area of λ, ϑ given by the above supposed intervals $\lambda \in \langle 0, 2\rangle$ and $\vartheta \in \langle 0.5, 3\rangle$. In these ranges the required damping ratio δ may be supposed as prescribed constant, $\delta \cong 0.35$ [13].

3.5 Windup-Observer Conditioning for the IMC Controller

In fact the real operation of the IMC controller can follow the state model formulation (3.13), (3.14) only as long as the actuating variable remains within the actuator range, $u_{\min} < u < u_{\max}$. As soon as any of the saturation boundaries is reached the saturated output cannot follow this model any more and becomes stuck at this boundary value until the applied anti-windup procedure restores the normal operation and this modification unavoidably leads to nonlinear nature of the control loop. We further need to distinguish between two controller outputs, namely between the actual saturated $u_S(\bar{t})$ and an auxiliary, internally estimated $\hat{u}(\bar{t})$. Similarly as in most of the referenced works the scheme of "state observer feedback" is applied for this estimation.

Definition 1 Consider a saturation free controller described by (3.13), (3.14), i.e. by the Laplace transform equations

$$\bar{s}x(\bar{s}) = Ax(\bar{s}) + Be(\bar{s}), \qquad u(\bar{s}) = Cx(\bar{s}) + De(\bar{s}), \tag{3.23}$$

with matrices given in (3.15). Let the system operating according to the following equations

$$\begin{aligned} \bar{s}\hat{x}(\bar{s}) &= A\hat{x}(\bar{s}) + Be(\bar{s}) + H[u_S(\bar{s}) - C\hat{x}(\bar{s}) - De(\bar{s})] \\ &= [A - HC]\hat{x}(\bar{s}) + [B - HD]e(\bar{s}) + Hu_S(\bar{s}), \end{aligned} \tag{3.24}$$

with a gain matrix $H = [h_1, h_2]^T$ be introduced as the *windup observer* where H is to be set by an anti-windup tuning procedure. The auxiliary output variable $\hat{u}(\bar{t})$ results from the state estimate $\hat{x}(\bar{t})$ and from the control error $e(\bar{t})$ as $\hat{u}(\bar{t}) = C\hat{x}(\bar{t}) + De(\bar{t})$.

The scheme given by (3.24) should not be confused with a genuine state observer. The observer-like feedback in Eq. (3.24) acts intermittently, being *switched on* and *off* in the instants of saturation. For instance the characteristic equation of (3.24)

$$\det[sI - A + HC] = \bar{s}^2 + (2\delta\Phi + h_2C_2)\bar{s} + (1+\delta^2)\Phi^2\,(1 - \exp(-\bar{s}) + h_1C_2 = 0. \tag{3.25}$$

$C_2 = (1 + \delta^2)\Phi^2/K$, contains terms originating from this feedback but due to the intermittent operation (3.25) cannot properly represent the actual dynamics of system (3.24) which is nonlinear in fact. Only intuitively we can reckon with a quicker observer response with increasing h_1, h_2. The state vector $\hat{x}(\bar{t})$ has a passing role only and with respect to its different purpose the feedback in (3.24) is used to be rearranged to a scheme with feedback closed from the controller output using the equality, $C\hat{x}(\bar{t}) + De(\bar{t}) = \hat{u}(\bar{t})$ given by the equation

$$\bar{s}\hat{x}(\bar{s}) = A(\bar{s})\hat{x}(\bar{s}) + B(\bar{s})e(\bar{s}) + H_s[u_s(\bar{s}) - \hat{u}(\bar{s})], \tag{3.26}$$

where the difference $\hat{u}(\bar{t}) - u_S(\bar{t})$ is further referred to as the *saturation error*. This scheme represents an inherent part of the controller itself. Whenever the saturation does not occur the identity $\hat{u}(\bar{t}) \equiv u_S(\bar{t})$ makes the last feedback term zero while during the saturation the error affects the dynamics of the state estimate $\hat{x}(\bar{t})$. Just switching *on* and *off* of this feedback makes the scheme (3.26) nonlinear and thus different from the genuine observer.

3.6 Tuning the Windup Observer to Optimize the Control Loop Response

The standard aim of tuning the anti-windup scheme is to minimize the time intervals when saturation affects the actuating variable, i.e. to minimize the saturation error $\hat{u}(\bar{t}) - u_S(\bar{t})$ [5]. i.e to keep the *actuating variable* as much as possible close to its ideal action. But (3.26) is an inherent part of the controller (3.11) and therefore the minimization of *control error* $e(\bar{t})$ is to be preferred to a strict minimizing the *saturation error*, $\hat{u} - u_S$. Therefore the optimization of the gains h_1, h_2 is made from the aspect that not an optimization of the *saturation error* $\hat{u}(\bar{t}) - u_S(\bar{t})$ but the best attainable performance *control error* $e(\bar{t})$ is searched in h_1, h_2 tuning.

The following proposition holds for tuning the feedback gain $H = [h_1,\ \ h_2]^T$.

Proposition 2 *Let the quality of control error $e(\bar{t})$ be evaluated by its absolute error integral (IAE). The symbols $\hat{e}(\bar{t})$ and $e_S(\bar{t})$ are introduced to distinguish between the saturation free and saturated alternatives of control loop performance respectively. For both of them the appropriate IAE functions are then as follows*

$$\hat{I}_A = \int_0^{\bar{t}} \left|\hat{e}(\sigma)\right| d\sigma, \; I_{AS} = \int_0^{\bar{t}} \left|e_S(\sigma)\right| d\sigma, \tag{3.27}$$

and their limits for $\bar{t} \to \infty$ are used as performance criteria $\hat{I}_A = \lim_{\bar{t}\to\infty} \hat{I}_A(\bar{t})$ and $I_{AS} = \lim_{\bar{t}\to\infty} I_{AS}(\bar{t})$. The ratio of them $R_{AE} = I_{AS}/\hat{I}_A$ is then considered as the *performance criterion* evaluating the impact of the windup observer (3.24) involvement in the control loop. The usual value $R_{AE} > 1$ indicates that the IAE performance becomes worse due to saturation, while $R_{AE} < 1$ indicates an improvement.

A novelty of the presented approach given by Proposition 2 consists in considering the plant properties in the tuning. However, it is easy to see that this way of setting the gains h_1, h_2 is feasible only due to the consistent use of the *dimensionless model* of the control loop. For a set of similar plants (3.3) with common λ, ϑ the IMC controller is designed according to (3.11) with specifying the filter dynamics by $\bar{p}_{1,2} = \Phi_k(-\delta \pm j)$. After applying (3.22) angle Φ_k can be eliminated and the whole IMC control loop is then completely identifiable only with the parameters λ, ϑ and the damping ratio δ. This property keeps hold even if controller (3.11) is rearranged to the state form (3.13), (3.14), (3.15). The involvement of the windup observer (3.24) is tuned by the gains h_1, h_2 which remain the only parameters to be optimized according to the R_{AE} ratio as the selected performance criterion. As to the damping ratio it was proved in [13] that for the dominance of the prescribed $\bar{p}_{1,2} = \Phi_k(-\delta \pm j)$ the *optimum value* is approximately $\delta \cong 0.35$. With this fixation only the numbers λ, ϑ identify each of the investigated options of the plant and IMC control loop.

For the class of IMC control loops given by a common option of λ, ϑ the best h_1, h_2 setting is then found as the minimum of $R_{AE}(h_1, h_2)$ over the $h_{1,2}$ area. This setting provides the saturating controller with the best attainable *control error* while the *saturation error* is primarily not regarded in criterion R_{AE}. In the next section we will learn the special radial shape of the criterion $R_{AE}(h_1, h_2)$ for a fixed λ, ϑ which makes it possible to respect *both* the errors $e(\bar{t})$ and $\hat{u} - u_S$ in applying the control error criterion $R_{AE}(h_1, h_2)$.

3.7 Application Example and the Tuning Rule

Consider the class of similar plants as in (3.1) where the parameters $\lambda = a_0/a_1^2 = 0.2$ and $\vartheta = a_1\tau = 2$. All these plants have the same ultimate angle given by (3.22) as $\Phi_k = 1.2647$ and therefore their IMC controller according to (3.11), with $\Phi = \Phi_k$ and the damping $\delta = 0.35$, results in the following common transfer function $R(\bar{s})$

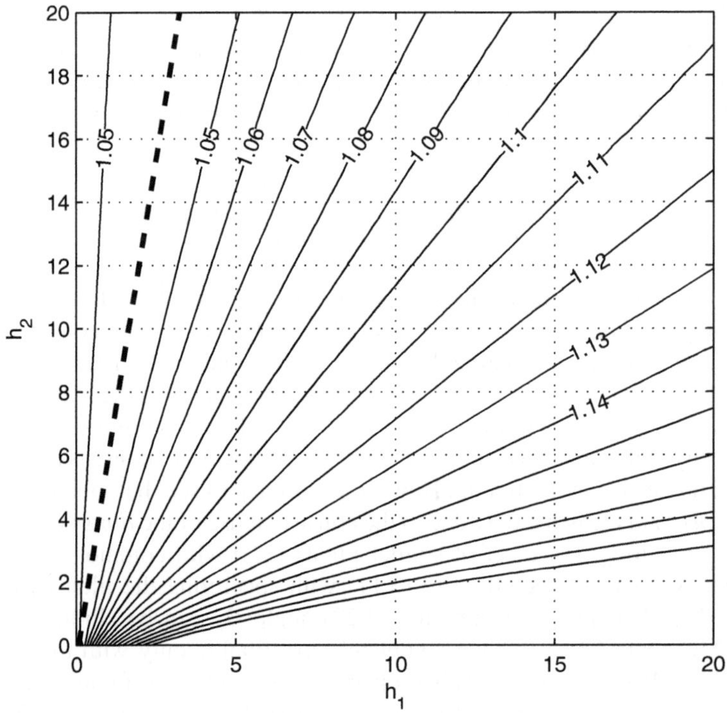

Fig. 3.2 Criterion R_{AE} values over area h_1, h_2 with optimum locus (*dashed line*)

$$R(\bar{s}) = \frac{1.7954(\bar{s}^2 + 2\bar{s} + 0.8)}{K\left[0.8\bar{s}^2 + 0.7082\bar{s} + 1.4363(1 - \exp(-\bar{s}))\right]}. \tag{3.28}$$

This controller is transformed into the state space form (3.13), (3.14), (3.15), and the windup observer feedback as in (3.24) is added. The control loop with saturation resulting from this controller and the considered plant may now be tested in repeated simulations on its *disturbance rejection* responses with various options of the gains h_1, h_2. For each of the simulation runs the criterion ratio $R_{AE}(h_1, h_2)$ is evaluated and the result of this testing over a sufficiently wide area of h_1, h_2 is in Fig. 3.2. The shape of the criterion $R_{AE}(h_1, h_2)$ is displayed by means of the contour lines $R_{AE} = const$ and the linear radial character of these lines reveals the following result: The $R_{AE}(h_1, h_2)$ optimum is not a specific setting point h_1, h_2 but a straight hollow so that it is provided by any pair h_1, h_2 satisfying a proportionality $h_2 = kh_1 + c$. For the presented example this proportionality is given by $k = 6.29, c = -0.41$ and is drawn by the dashed line in Fig. 3.2. In view of this property in evaluating the criterion $R_{AE}(h_1, h_2)$ it is possible to take into account not only the optimum of *control error* but also the least obtainable *saturation error*. Apparently the higher the gains h_1, h_2 the lower the saturation error is obtained. Therefore if any of the

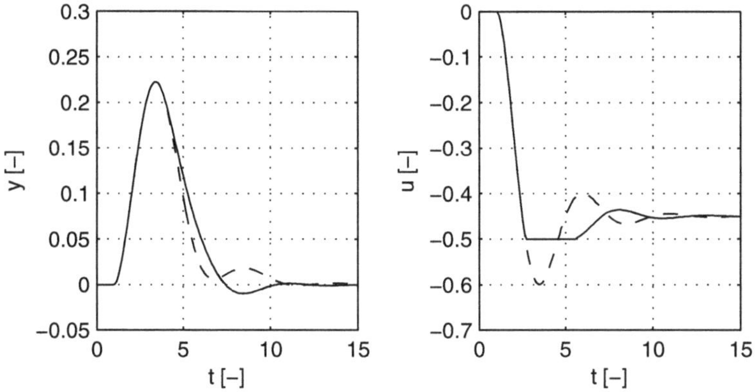

Fig. 3.3 Disturbance rejection response obtained with controller (3.28) with (*solid line*) and without saturation (*dashed line*)

settings satisfying $h_2 = kh_1 + c$ are optimum as to $R_{AE}(h_1, h_2)$, then the maximum values of h_1, h_2 satisfying this relation provide an optimum for both the control and the saturation errors respectively. Hence the recommended setting in this example is $h_1 = 3.24, h_2 = 20$. The step responses of the disturbance rejection with and without saturation are in Fig. 3.3. The selection of as high values of the gains is explained below.

From the extensive set of simulation experiments it has resulted that the above demonstrated optimum achieved by a proportionality between h_1 and h_2 is not an exception but a *rule*. The radial shape of $R_{AE}(h_1, h_2)$ criterion is a feature that is characteristic for the plants (3.3) in general, only the constants k and c change with varying λ and ϑ. For any plant option given by λ and ϑ one can identify the corresponding values of k and c. The straight lines $h_2 = kh_1 + c$, as in Fig. 3.2 determine the $R_{AE}(h_1, h_2)$ minimum by almost all of their points with the exception of an area around the h_1, h_2 origin. In fact the higher $|h_{1,2}|$ the better the appropriate part of the straight line describes the minimum. Besides, the higher values of h_1, h_2 the quicker is the compensation of the saturation error. That is why the highest h_1, h_2 values from the obtained optimum line are taken. During the numerous experiments with h_1, h_2 variations it has been found that increasing the gains over 20, approximately, is inefficient in the observer performance. That is why this value is chosen as the upper bound of both h_1, h_2.

After investigating a representative set of plants from the ranges $\lambda \in \langle 0, 1.5\rangle$ and $\vartheta \in \langle 0.5, 3\rangle$ and evaluating their $R_{AE}(h_1, h_2)$ criterion the appropriate proportionality constants k and c can be assessed for each of the pairs λ, ϑ. The obtained results are presented in Table 3.1.

Table 3.1 Values of the coefficients k, c for selected λ, ϑ combinations

k	ϑ						c	ϑ					
	0.5	0.7	1.0	1.5	2.0	2.5		0.5	0.7	1.0	1.5	2.0	2.5
λ 0.1	91.2	91.2	91.2	91.2	91.2	91.2	0.1	0.29	0.29	0.29	0.29	0.29	0.29
0.2	91.2	91.2	91.2	91.2	6.3	2.7	0.2	0.29	0.29	0.29	0.29	−0.41	−1.33
0.3	91.2	15.5	4.3	1.3	0.6	0.3	0.3	0.29	1.08	−0.44	−0.66	−0.77	−0.42
0.5	2.5	1.0	0.4	0.2	0.1	0.1	0.5	−0.35	−0.27	0.07	−0.46	−0.60	−0.47
1.0	0.2	0.1	0	0	0.1	0.2	1.0	0.57	−0.60	0	−0.22	0.55	1.87
1.5	0.1	0	0	0	0.2	0.4	1.5	−0.30	0	0	0.40	3.54	4.83

3.8 Concluding Remarks

The IMC control loop with a meromorphic controller is investigated from the point of view of the actuator saturation. It has been demonstrated that the usual observer-like scheme is applicable to the meromorphic controller function too. Unlike the commonly used approach to the windup problem the actuator saturation is not regarded as a separate nonlinearity but as an inseparable property of the controller. From this point of view the *control error* rather than *saturation error* is preferred in tuning the anti-windup scheme parameters. Although the windup observer feedback is closed from the saturation error its tuning is performed primarily subject to the IAE control error performance criterion. It is necessary to underline that this approach was possible only due to the dimensionless description of the whole IMC control loop, i.e. by the use of the *swingability* and *laggardness* similarity numbers. As it results from the Buckinghams theorem also the h_1, h_2 tuning is dependent only on these similarity numbers. During the whole chapter the *disturbance rejection* point of view is regarded as decisive for the control loops with a substantial delay.

Acknowledgments The presented research was supported by The Technology Agency of the Czech Republic under the Competence Centre Project TE01020197, Centre for Applied Cybernetics 3.

References

1. Fertik, H.A., Ross, C.W.: Direct digital control algorithm with anti-windup feature. ISA Trans. **6**(4), 317–328 (1967)
2. Åström, K.J., Rundquist, L.: Integrator windup and how to avoid it. In: American Control Conference, 1989. IEEE, pp. 1693–1698 (1989)
3. Kapoor, N., Teel, A.R., Dautidis, P.: An anti-windup design for linear systems with input saturation. Automatica **34**(5), 559–574 (1998)
4. Kothare, M.V., et al.: A unified framework for the study of anti-windup designs. Automatica **30**(12), 1869–1883 (1994)
5. Hanus, R., Kinnaert, M., Henriotte, J.-L.: Conditioning technique, a general anti-windup and bumpless transfer method. Automatica **23**(6), 729–739 (1987)

6. Doyle, J.C., Smith, R.S., Enns, D.F.: Control of plants with input saturation nonlinearities. In: American Control Conference, 1987. IEEE, pp. 1034–1039 (1987)
7. Edwards, C., Poslethwaite, I.: Anti-windup and bumpless-transfer schemes. Automatica **34**(2), 199–210 (1998)
8. Weston, P.F., Postlethwaite, I.: Linear conditioning for systems containing saturating actuators. Automatica **36**(9), 1347–1354 (2000)
9. Turner, M.C., Postlethwaite, I.: A new perspective on static and low order anti-windup synthesis. Int. J. Control **77**(1), 27–44 (2004)
10. Turner, M.C., Herrmann, G., Postlethwaite, I.: Incorporating robustness requirements into anti-windup design. IEEE Trans. Autom. Control **52**(10), 1842–1855 (2007)
11. Zítek, P., Kučera, V.: Algebraic design of anisochronic controllers for time delay systems. Int. J. Control **76**, 1654–1665 (2003)
12. Zheng, A., Kothare, M.V., Morari, M.: Anti-windup design for internal model control. Int. J. Control **60**(5), 1015–1024 (1994)
13. Zítek, P., Fišer, J., Vyhlídal, T.: Dimensional analysis approach to dominant three-pole placement in delayed PID control loops. J. Process Control **23**(8), 1063–1074 (2013)
14. Goodwin, G.C., Graebe, S.F., Salgado, M.E.: Control System Design. Prentice Hall Inc, Upper Saddle River (2001)

Chapter 4
Stabilization of Some Fractional Neutral Delay Systems Which Possibly Possess an Infinite Number of Unstable Poles

Le Ha Vy Nguyen and Catherine Bonnet

Abstract We consider fractional delay systems of neutral type and prove that many systems with infinitely many unstable poles cannot be stabilized by the class of rational fractional controllers of commensurate order. For a class of fractional neutral delay systems with an infinite number of poles asymptotic to the imaginary axis from the right or left hand side, we are able to derive a parametrization of all stabilizing controllers from fractional PI controllers obtained from previous work.

4.1 Introduction

Fractional systems are systems involving non-integer derivatives in the time domain and thus containing power of non-integer order of the Laplace variable s (i.e. s^{μ} with μ non-integer) in the frequency domain. Such models appear in many engineering fields since they describe certain systems better than their integer counterpart, see for example [11] and references therein.

As delays are usually encountered in real-life situations, fractional systems with delays have been of increasing interests in the past ten years. Stability of this kind of systems has been studied in several publications such as [1–5, 7, 12, 13, 16]. However, not many results are available for stabilization. Some references are [4] on H_{∞}-stabilization , [19] on sliding mode control, [9, 10] on stabilization of dead-time fractional systems, and [14] on finite-time stabilization.

In [4], a rather complete study from stability to stabilization is realized for a class of neutral fractional systems with one delay. The main result of this study is a parametrization of all stabilizing controllers. However, the problem of finding a

L. H. V. Nguyen (✉) · C. Bonnet
Inria, LSS-SUPELEC, 3 rue Joliot Curie, 91192 Gif-sur-Yvette cedex, France
e-mail: Le-Ha-Vy.Nguyen@inria.fr

C. Bonnet
e-mail: Catherine.Bonnet@inria.fr

A. Seuret et al. (eds.), *Low-Complexity Controllers for Time-Delay Systems*,
Advances in Delays and Dynamics 2, DOI: 10.1007/978-3-319-05576-3_4,

parametrization for systems with large poles clustering the imaginary axis remained unsolved. Later, the stability of neutral fractional systems with commensurate delays and with large poles asymptotic to the imaginary axis has been studied in [6]. In this chapter we apply these stability results to examine some stabilizability properties of fractional neutral systems with commensurate delays and an infinite number of unstable poles. We find that a majority of these systems cannot be stabilized by the class of rational fractional controllers of commensurate order except systems with the lowest degree. The systems considered in [4] belongs to this exception. A parametrization of stabilizing controllers is derived for these systems.

Remark that some systems we consider (those with infinitely many unstable poles) belong to the class considered in [8]. However the parametrization of all stabilizing controllers they proposed requires an inner/outer factorization of the plant.

The rest of the chapter is organised as follows. In Sect. 4.2, we examine the stabilizability of neutral fractional systems with commensurate delays and with chains of poles in the right half-plane. In Sect. 4.3, we obtain a parametrization of all stabilizing controllers for systems with only one delay, allowing here chains of poles clustering the imaginary axis from the left or the right hand side. Finally, we conclude the chapter by Sect. 4.4.

4.2 Stabilizability Properties of Fractional Systems with Commensurate Delays

We study the H_∞-stabilization of fractional time-delay systems of neutral type with commensurate delays given as follows

$$G(s) = \frac{1}{p(s) + \sum_{k=1}^{N} q_k(s)e^{-skh}}, \tag{4.1}$$

where $h > 0$ is the delay, p and q_k, $k \in \mathbb{N}_N$[1] are real polynomials in s^μ, $\mu \in (0, 1)$, $\deg p \geq 1$, $\deg p \geq \deg q_k$, and there is at least one q_k, $k \in \mathbb{N}_N$ such that $\deg p = \deg q_k$. Here the degree is interpreted as the degree in s^μ and so is an integer. In order to avoid multi-valued functions, we consider s in the Riemann sheet such that $-\pi < \arg s < \pi$.

Since $\deg p \geq \deg q_k$ for all $k \in \mathbb{N}_N$, we can obtain the following expression for each k

$$\frac{q_k(s^\mu)}{p(s^\mu)} = \alpha_k + \frac{\beta_k}{s^\mu} + \frac{\gamma_k}{s^{2\mu}} + \mathcal{O}(s^{-3\mu}) \text{ as } |s| \to \infty,$$

[1] We denote $\mathbb{N}_N = \{1, \ldots, N\}$.

where

$$\alpha_k = \frac{\sigma_{k,d}}{\rho_d},$$
$$\beta_k = \frac{\sigma_{k,d-1} - \alpha_k \rho_{d-1}}{\rho_d},$$

d denotes the degree of p in s^μ, $p(s) = \sum_{m=0}^{d} \rho_m s^{\mu m}$ and $q_k(s) = \sum_{m=0}^{d} \sigma_{k,m} s^{\mu m}$.

The coefficient of the highest degree term of the denominator of the transfer function (4.1) can be written as a multiple of the following polynomial (called formal polynomial) in z

$$\widetilde{c}_d(z) = 1 + \sum_{k=1}^{N} \alpha_k z^k, \tag{4.2}$$

where $z = e^{-sh}$.

According to Proposition 2.1 in [6], each neutral chain of poles of G is associated to each root r of (4.2) and is first approximated by

$$s_n h = \lambda_n + o(1),$$

where

$$\lambda_n = -\ln(r) + j2\pi n, \quad n \in \mathbb{Z}.$$

As a consequence, the neutral chain of poles asymptotically approaches the vertical line

$$\Re(s) = -\frac{\ln(|r|)}{h}. \tag{4.3}$$

Let us consider controllers of the form

$$K(s) = \frac{N(s)}{D(s)}, \tag{4.4}$$

where N and D are real polynomials in s^μ, $\mu \in (0, 1)$. These controllers are called rational fractional controllers of commensurate order. From Lemma 4.1 in [17], we know that $\deg N \leq \deg D$ if $K(s)$ stabilizes $G(s)$ in the sense H_∞. Suppose that $N(s)$ and $D(s)$ do not have common zeros, and $N(s)$ does not have common zeros with the denominator of $G(s)$.

The closed-loop $[G, K]$ is stable if and only if the following transfer functions are stable

$$\frac{1}{1 + G(s)K(s)} = \frac{D(s)(p(s) + \sum_{k=1}^{N} q_k(s)e^{-skh})}{D(s)(p(s) + \sum_{k=1}^{N} q_k(s)e^{-skh}) + N(s)}, \tag{4.5}$$

$$\frac{G(s)}{1+G(s)K(s)} = \frac{D(s)}{D(s)(p(s)+\sum_{k=1}^{N} q_k(s)e^{-skh})+N(s)}, \tag{4.6}$$

$$\frac{K(s)}{1+G(s)K(s)} = \frac{N(s)(p(s)+\sum_{k=1}^{N} q_k(s)e^{-skh})}{D(s)(p(s)+\sum_{k=1}^{N} q_k(s)e^{-skh})+N(s)}. \tag{4.7}$$

Under the assumptions about the zeros of $N(s)$ and $D(s)$, the transfer functions (4.5), (4.6), and (4.7) do not have zero cancellation between the numerator and the denominator.

If $\widetilde{c}_d(z)$ has a root r with $|r| < 1$, then due to (4.3), the chain of poles relative to r is asymptotic to a vertical line lying in the open right half-plane. Thus this chain has infinitely many poles in the open right half-plane and the system is unstable. The stabilization of such systems under controllers of the form (4.4) is examined in the following proposition.

Proposition 1 *Let G be given as in* (4.1). *If $\widetilde{c}_d(z)$ has roots of modulus strictly smaller than one, then G cannot be stabilized by a controller given as in* (4.4).

Proof Since $\deg N \leq \deg D$ and $\deg p \geq 1$, the denominator of the closed-loop transfer function (4.6) also has the formal polynomial $\widetilde{c}_d(z)$ with roots of modulus strictly smaller than one. □

Another situation where G has an infinite number of poles in the open right half-plane is when G has chains of neutral poles approaching the imaginary axis from the right and the other neutral chains asymptotic to vertical lines in the open left half-plane. The next proposition addresses this class of systems.

Proposition 2 *Let G be given by* (4.1). *Suppose that the polynomial $\widetilde{c}_d(z)$ has roots of modulus one of multiplicity one and that the other roots are of modulus greater than one. Suppose also that at least one root of modulus one of $\widetilde{c}_d(z)$, denoted r, satisfies*

$$\Re(\delta_n) > 0 \tag{4.8}$$

where

$$\delta_n = \frac{h^{\mu}\sum_{k=1}^{N}\beta_k r^k}{(2jn\pi)^{\mu}\sum_{k=1}^{N}k\alpha_k r^k}.$$

Then G can be stabilized by controllers of the form (4.4) *only if* $\deg p = 1$.

Proof Recall from Theorem 2.1 in [6] that the poles of large modulus corresponding to r, denoted by s_n, are approximated by

$$s_n h = -\ln(r) + 2jn\pi + \delta_n + \mathcal{O}(n^{-2\mu}).$$

Under the assumption (4.8), we see that G has infinitely many poles in the open right half-plane.

Let us examine the denominator of the transfer functions of the closed-loop which is

$$D(s)(p(s)+\sum_{k=1}^{N} q_k(s)e^{-skh})+N(s)=D(s)p(s)+N(s)+\sum_{k=1}^{N} D(s)q_k(s)e^{-skh}.$$

We consider the development at infinity of

$$\frac{D(s)q_k(s)}{D(s)p(s)+N(s)}=\bar{\alpha}_k+\frac{\bar{\beta}_k}{s^{\mu}}+o(s^{-\mu}).$$

As $\deg p \geq 1$ and $\deg D \geq \deg N$, we have that $\bar{\alpha}_k=\alpha_k$. Now, if $\deg p > 1$, we also have that $\bar{\beta}_k=\beta_k$. In this case, the closed-loop has an infinite number of unstable poles and thus cannot be H_∞-stable. □

Remark 1 The systems considered in Proposition 2 are not the only ones with chains of poles approaching the imaginary axis from the right. This may also happen to systems with $\Re(\delta_n)=0$ but this case needs further analysis as described in [5].

4.3 Parametrization of the Set of Stabilizing Controllers in a Particular Case

The simplest systems described by (4.1) and with $\deg p = 1$ are systems with one delay. They have been studied in [4]. For such systems with transfer function given by

$$G(s)=\frac{1}{(as^{\mu}+b)+(cs^{\mu}+d)e^{-sh}},\tag{4.9}$$

where $a, b, c, d \in \mathbb{R}$, $a > 0$, $|a|=|c|$, and $\mu \in (0,1)$, fractional PI controllers have been obtained.

These controllers are the starting point to obtain a parametrization of all stabilizing controllers, which is the main result of this section. Before stating the main result, we will recall the results on fractional PI controllers in [4]. First, to simplify its presentation, we derive the opposite condition to Remark 4.1 in [4].

Lemma 1 *$a_2z^2+a_1z+a_0=0$ with $a_2, a_1, a_0 \in \mathbb{R}$, $a_2 > 0$ has all roots in $\{z \in \mathbb{C}\backslash\{0\} : |\mathrm{Arg}(z)| > \mu\pi/2\}$ with $\mu \in (0,1)$ if and only if $a_0 > 0$ and $a_1 > -2\sqrt{a_0a_2}$ $\cos(\mu\pi/2)$.*

Proof The equation has two strictly negative roots if and only if

$$\begin{cases} \Delta = a_1^2 - 4a_0a_2 \geq 0 \\ a_1 > 0 \\ a_0 > 0 \end{cases}$$

$$\Leftrightarrow \begin{cases} \Delta = (a_1 - 2\sqrt{a_0a_2})(a_1 + 2\sqrt{a_0a_2}) \geq 0 \\ a_1 > 0 \\ a_0 > 0 \end{cases}$$

$$\Leftrightarrow \begin{cases} a_1 \geq 2\sqrt{a_0a_2} \\ a_0 > 0 \end{cases}$$

The equation has two complex conjugate roots, denoted $re^{\pm j\phi}$, with $\phi \in (\mu\pi/2, \pi) \cup (-\pi, -\mu\pi/2)$ if and only if

$$\begin{cases} \Delta = a_1^2 - 4a_0a_2 < 0 \\ a_0 > 0 \\ \cos\phi = -\frac{a_1}{2\sqrt{a_0a_2}} < \cos\left(\frac{\mu\pi}{2}\right) \end{cases}$$

$$\Leftrightarrow \begin{cases} a_1 < 2\sqrt{a_0a_2} \\ a_0 > 0 \\ a_1 > -2\sqrt{a_0a_2}\cos(\mu\pi/2) \end{cases}$$

□

We now recall the characterization of H_∞-stabilizing fractional PI controllers of systems (4.9).

Proposition 3 **(Proposition 4.1** [4]) *Let G be given by* (4.9) *and $K(s) = k_p + k_i/s^\mu$ with $k_p, k_i \in \mathbb{R}$.*

1. *Let $a = c$. If k_p and k_i satisfy $\frac{b+d+k_p}{a+c} > -2\sqrt{\frac{k_i}{a+c}}\cos\left(\frac{\mu\pi}{2}\right)$ and $k_i > 0$ then K stabilizes G when $h = 0$.*
   Moreover, if $a(b + k_p - d)\cos(\frac{\mu\pi}{2}) > 0$, then K stabilizes G for small h.
   If k_p and k_i satisfy also $(b + k_p)^2 + 2ak_i\cos(\mu\pi) - d^2 > 0$ and $k_i(b + k_p)\cos(\frac{\mu\pi}{2}) > 0$ then K stabilizes G for all h.
2. *Let $a = -c$. If $k_i(b + k_p + d) > 0$, then K stabilizes G when $h = 0$.*
   Moreover, if $a(b + k_p + d)\cos(\frac{\mu\pi}{2}) > 0$ then K stabilizes G for small h.
   If k_p and k_i satisfy also $(b + k_p)^2 + 2ak_i\cos(\mu\pi) - d^2 > 0$ and $k_i(b + k_p)\cos(\frac{\mu\pi}{2}) > 0$ then K stabilizes G for all h.

Remark 2 The system G considered in Proposition 3 may have infinitely many poles in the open right or left half-plane.

In the following examples, we observe the change of side of the chain of poles from open-loop to closed-loop. We have $a = c$ for the first example and $a = -c$ for the second.

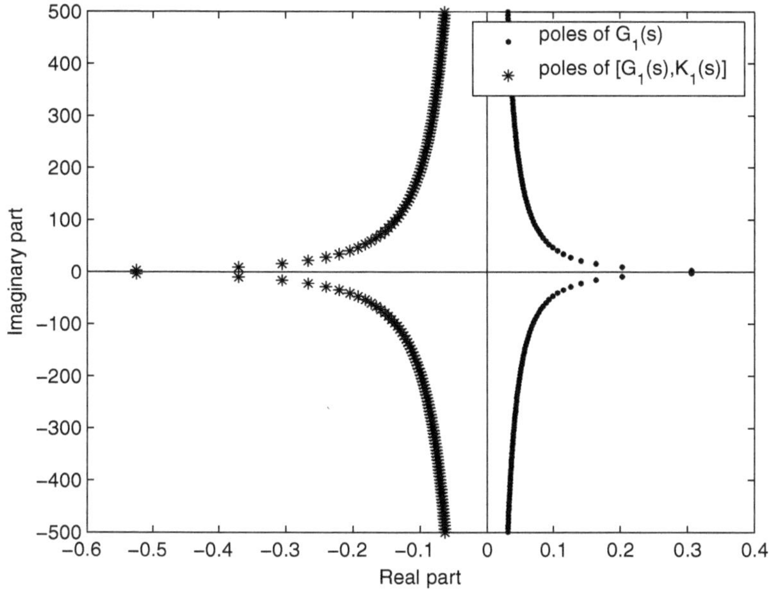

Fig. 4.1 Poles of $G_1(s)$ and of the closed-loop system $[G_1(s), K_1(s)]$

Example 1

$$G_1(s) = \frac{1}{(s^{1/2}+1)+(s^{1/2}+2)e^{-s}},$$
$$K_1(s) = 3 + \frac{2}{s^{1/2}}. \tag{4.10}$$

Example 2

$$G_2(s) = \frac{1}{(s^{1/2}-3)+(-s^{1/2}+1)e^{-s}},$$
$$K_2(s) = 5 + \frac{1}{s^{1/2}}.$$

The poles of the open-loop and of the closed-loop systems in Example 1 and Example 2 are computed by QPmR algorithm [21] and are showed in Fig. 4.1 and Fig. 4.2 respectively.

Given $K_0(s)$, a stabilizing controller of the system (4.9), we can directly obtain a parametrization of all stabilizing controllers without finding coprime factorizations by using Theorem 2 in [18].

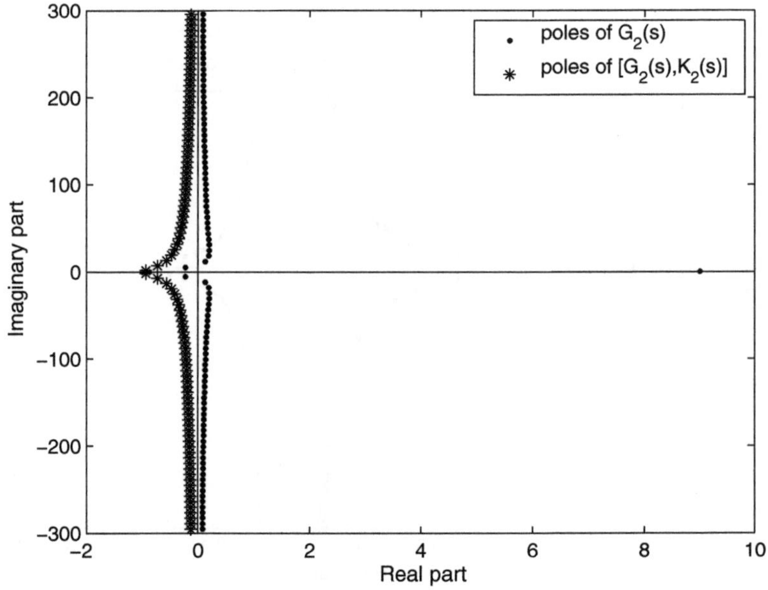

Fig. 4.2 Poles of $G_2(s)$ and of the closed-loop system $[G_2(s), K_2(s)]$

Proposition 4 *Let $G(s)$ be given as in* (4.9). *A parametrization with two degrees of freedom of all H_∞-stabilizing controllers of $G(s)$ is given by*

$$\frac{-T(s^\mu R+T)+(s^{2\mu}Q_1+Q_2T^2)R}{s^\mu(s^\mu R+T)+(s^{2\mu}Q_1+Q_2T^2)}, \tag{4.11}$$

where $Q_1, Q_2 \in H_\infty$ are two free parameters,

$$R(s) = (as^\mu + b) + (cs^\mu + d)e^{-sh},$$
$$T(s) = k_p s^\mu + k_i,$$

$k_i > 0$ and k_p satisfy

$$\frac{b+d+k_p}{a+c} > -2\sqrt{\frac{k_i}{a+c}}\cos\left(\frac{\mu\pi}{2}\right) \quad \text{for } a = c,$$
$$k_i(b+d+k_p) > 0 \quad \text{for } a = -c,$$
$$(a(b+k_p) - cd)\cos\left(\frac{\mu\pi}{2}\right) > 0,$$
$$(b+k_p)^2 + 2ak_i\cos(\mu\pi) - d^2 > 0,$$
$$k_i(b+k_p)\cos\left(\frac{\mu\pi}{2}\right) > 0.$$

Proof From Proposition 3, we have that the fractional PI controller given by $K_0(s) = k_p + k_i/s^\mu$ stabilizes $G(s)$. Denoting

$$A(s) = \frac{1}{1 + G(s)K_0(s)},$$
$$B(s) = -\frac{K_0(s)}{1 + G(s)K_0(s)},$$

we have that $A(s)$ and $B(s)$ satisfy $0 \neq A(s), B(s) \in H_\infty$, $A(s) - B(s)G(s) = 1$ and $A(s)G(s) \in H_\infty$. Then Theorem 2 in [18] shows that a parametrization of all stabilizing controllers of $G(s)$ in the sense of H_∞ is given by

$$\frac{B + Q_1A^2 + Q_2B^2}{A + Q_1A^2G + Q_2B^2G}$$

where $Q_1, Q_2 \in H_\infty$ are two free parameters. This gives (4.11). □

However, the Youla-Kučera parametrization (with one parameter) may be more favourable for controller design and as G is H_∞-stabilizable, we know that G necessarily admits a coprime factorization over H_∞ [20]. The next proposition, which characterizes quasi-polynomials with all roots in the open left half-plane, will be useful to find a coprime factorization of the transfer function of the system, which is required to derive a parametrization of stabilizing controllers using Youla-Kučera formula.

Proposition 5 *The equation*

$$(as^\mu + b) + (cs^\mu + d)e^{-sh} = 0, \tag{4.12}$$

with $a > 0$, $b, c, d, h \in \mathbb{R}$, $|a| = |c|$, and $\mu \in (0, 1)$ has no roots in the closed right half-plane for all $h \geq 0$ if and only if

- $b + d > 0$ *and* $b - d > 0$ *if* $a = c$,
- $b + d > 0$ *and* $b - d \geq 0$ *if* $a = -c$.

Proof From [15], we have that the equation has no roots in the closed right half-plane for all h if and only if

- the equation has no roots in the closed right half-plane for $h = 0$,
- infinitely many poles approach the imaginary axis from the left side for h sufficiently small,
- there is no roots crossing the imaginary axis for $h > 0$.

First, we consider the case $a = c$.

When $h = 0$, (4.12) becomes

$$2as^\mu + b + d = 0,$$

which has no root in the closed right half-plane if and only if $b + d > 0$.

For $h > 0$, no crossings means $W(\omega) := |p(i\omega)|^2 - |q(i\omega)|^2 \neq 0 \ \forall \omega > 0$ [4]. We have

$$W(\omega) = (a^2 - c^2)\omega^{2\mu} + b^2 - d^2 + 2\omega^{\mu} \cos\left(\frac{\mu\pi}{2}\right)(ab - cd). \tag{4.13}$$

For $a = c$, the above expression becomes

$$W(\omega) = (b - d)\left[(b + d) + 2a\omega^{\mu} \cos\left(\frac{\mu\pi}{2}\right)\right].$$

Then $W(\omega) \neq 0 \ \forall \omega > 0$ if and only if $b \neq d$ and $b + d \geq 0$.

Next, we have

$$\frac{as^{\mu} + b}{cs^{\mu} + d} = \frac{a}{c} + \frac{bc - ad}{c^2}\frac{1}{s^{\mu}} + \mathscr{O}(s^{-2\mu}).$$

Since $bc - ad \neq 0$ from the previous argument, Theorem 3.1 in [4] shows that large roots of (4.12) are stable if and only if

$$\begin{aligned} &\frac{bc - ad}{c^2}\frac{c}{a} > 0 \\ &\Leftrightarrow b - d > 0 \end{aligned} \tag{4.14}$$

From the three conditions, we derive that $b + d > 0$ and $b - d > 0$.

Similarly, we consider the case $a = -c$.

For $h = 0$, (4.12) becomes

$$b + d = 0,$$

which has no root in the closed right half-plane if and only if $b + d \neq 0$.

For $a = -c$, (4.13) becomes

$$W(\omega) = (b + d)\left[(b - d) + 2a\omega^{\mu} \cos\left(\frac{\mu\pi}{2}\right)\right].$$

Then $W(\omega) \neq 0 \ \forall \omega > 0$ if and only if $b + d \neq 0$ and $b - d \geq 0$.

With $a = -c > 0$, the condition (4.14) for stable chain of poles becomes $b + d > 0$.

From the three conditions, we derive that $b + d > 0$ and $b - d \geq 0$. □

Now we give a parametrization of all stabilizing controllers of systems given by (4.9).

Theorem 1 *Let*

$$G(s) = \frac{1}{(as^{\mu} + b) + (cs^{\mu} + d)e^{-sh}},$$

with $a, b, c, d \in \mathbb{R}$, $a > 0$, $|a| = |c|$, and $\mu \in (0, 1)$. The set of all H_∞-stabilizing controllers is given by

$$\frac{V + MQ}{U - NQ},$$

where

$$N(s) = \frac{1}{(a's^\mu + b') + (c's^\mu + d')e^{-sh}},$$

$$M(s) = \frac{(as^\mu + b) + (cs^\mu + d)e^{-sh}}{(a's^\mu + b') + (c's^\mu + d')e^{-sh}},$$

$$U(s) = \frac{s^\mu[(a's^\mu + b') + (c's^\mu + d')e^{-sh}]}{s^\mu(as^\mu + b + k_p) + k_i + s^\mu(cs^\mu + d)e^{-sh}}, \tag{4.15}$$

$$V(s) = \frac{(k_p s^\mu + k_i)[(a's^\mu + b') + (c's^\mu + d')e^{-sh}]}{s^\mu(as^\mu + b + k_p) + k_i + s^\mu(cs^\mu + d)e^{-sh}}, \tag{4.16}$$

Q is a free parameter in H_∞, $k_i > 0$ and k_p satisfy

$$\frac{b + d + k_p}{a + c} > -2\sqrt{\frac{k_i}{a + c}}\cos\left(\frac{\mu\pi}{2}\right) \quad \text{for } a = c,$$
$$k_i(b + d + k_p) > 0 \quad \text{for } a = -c,$$
$$(a(b + k_p) - cd)\cos\left(\frac{\mu\pi}{2}\right) > 0,$$
$$(b + k_p)^2 + 2ak_i\cos(\mu\pi) - d^2 > 0,$$
$$k_i(b + k_p)\cos\left(\frac{\mu\pi}{2}\right) > 0,$$

and $a', b', c', d' \in \mathbb{R}$ satisfy

$$a' > 0,$$
$$\frac{a'}{c'} = \frac{a}{c},$$
$$b' + d' > 0, \tag{4.17}$$
$$b' - d' > 0. \tag{4.18}$$

Proof Under the above conditions, $(a's^\mu + b') + (c's^\mu + d')e^{-sh}$ has no poles in the closed right half-plane. Now, $M(s)$ can be decomposed as follows

$$M(s) = \frac{a}{a'} + \frac{b - \frac{a}{a'}b'}{(a's^\mu + b') + (c's^\mu + d')e^{-sh}} + \frac{d - \frac{a}{a'}d'}{(a's^\mu + b') + (c's^\mu + d')e^{-sh}}e^{-sh}.$$

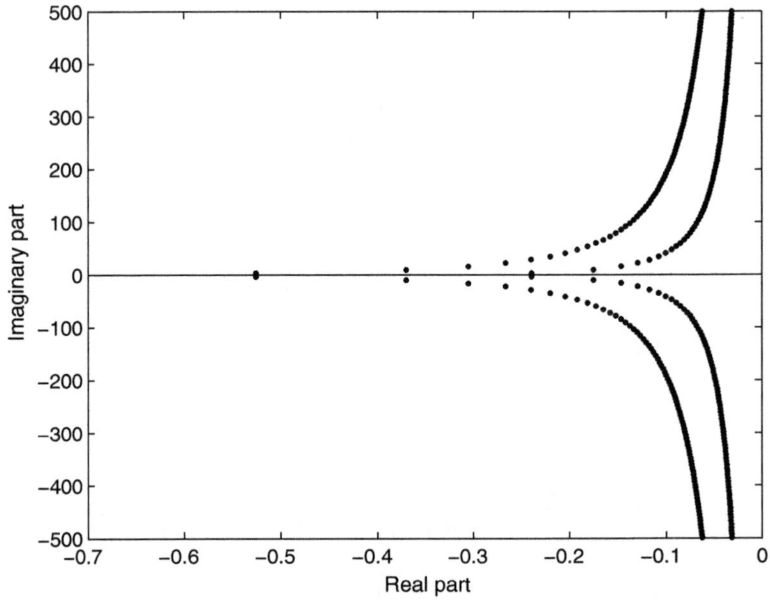

Fig. 4.3 Poles of the transfer functions of the closed-loop system $[G_1(s), \bar{K}_1(s)]$

Under the conditions (4.17) and (4.18), Corollary 3.2 in [4] shows that $N(s)$ and $M(s)$ belong to H_∞.

It is also easy to see that $\inf_{\Re(s)>0}(|N(s)| + |M(s)|) > 0$ so that (N, M) is a coprime factorization of G over H_∞.

By the same arguments as in [4], knowing a stabilizing controller K_0, one can derive the pair of Bézout factors U, V from the following expressions

$$\frac{1}{1+GK_0} = MU,$$
$$\frac{K_0}{1+GK_0} = MV.$$

Now, U and V in (4.15), (4.16) are obtained by using a PI controller proposed in Proposition 3. By decomposing U and V as in [4], we conclude that $U, V \in H_\infty$. □

The following example shows another stabilizing controller of the system G_1 in Example 1 obtained from the above parametrization.

Example 3 A coprime factorization of G_1 is

$$N(s) = \frac{1}{(s^{1/2}+3) + (s^{1/2}+2)e^{-s}},$$
$$M(s) = \frac{(s^{1/2}+1) + (s^{1/2}+2)e^{-s}}{(s^{1/2}+3) + (s^{1/2}+2)e^{-s}}.$$

Besides, $U(s)$ and $V(s)$ are obtained based on the PI controller as in (4.10).

Now, in order to have another controller we choose $Q \in H_\infty$. The simplest case is a constant, e.g. $Q = 1$. The corresponding controller is

$$\begin{aligned}\bar{K}_1(s) &= \frac{V(s)+M(s)}{U(s)-N(s)} \\ &= [4s^{3/2}+25s+45s^{1/2}+20+(8s^{3/2}+43s+68s^{1/2}+28)e^{-s} \\ &\quad +(4s^{3/2}+18s+24s^{1/2}+8)e^{-2s}]/[s^{3/2}+5s+5s^{1/2}-2 \\ &\quad +(2s^{3/2}+9s+10s^{1/2})e^{-s}+(s^{3/2}+4s+4s^{1/2})e^{-2s}]\end{aligned}$$

which involves commensurate delays. The poles of the closed-loop system are showed in Fig. 4.3.

4.4 Conclusion

In this chapter, we have considered first the stabilization of general fractional delay systems of the neutral type by rational fractional controllers of commensurate order. Then, for the special class of fractional delay systems with one delay we have derived a two-degree-of-freedom parametrization of the set of all H_∞-stabilizing controllers and given an explicit expression of coprime and Bézout factors allowing to obtain the Youla-Kučera parametrization of all the H_∞-stabilizing controllers. Future work will be devoted to the stabilization of a larger class of systems with several delays.

Acknowledgments We thank Dr. Alban Quadrat for many helpful discussions.

References

1. Akbari Moornani, K., Haeri, M.: On robust stability of LTI fractional-order delay systems of retarded and neutral type. Automatica **46**, 362–368 (2010)
2. Akbari Moornani, K., Haeri, M.: Necessary and sufficient conditions for BIBO-stability of some fractional delay systems of neutral type. IEEE Trans. Autom. Control **56**, 125–128 (2011)
3. Bonnet, C., Partington, J.R.: Analysis of fractional delay systems of retarded and neutral type. Automatica **38**, 1133–1138 (2002)
4. Bonnet, C., Partington, J.R.: Stabilization of some fractional delay systems of neutral type. Automatica **43**, 2047–2053 (2007)
5. Bonnet, C., Fioravanti, A.R., Partington, J.R.: Stability of neutral systems with commensurate delays and poles asymptotic to the imaginary axis. SIAM J. Control Optim. **49**, 498–516 (2011)
6. Fioravanti, A.R., Bonnet, C., Özbay, H.: Stability of fractional neutral systems with multiple delays and poles asymptotic to the imaginary axis. In: 49th IEEE Conference on Decision and Control, Atlanta, USA (2010)
7. Fioravanti, A.R., Bonnet, C., Özbay, H., Niculescu, S.-I.: Stability windows and unstable root-loci for linear fractional time-delay systems. In: Proceedings of the 18th IFAC World Congress, Milan, Italy (2011)

8. Gümüşsoy, S., Özbay, H.: On the mixed sensitivity minimization for systems with infinitely many unstable modes. Syst. Control Lett. **53**, 211–216 (2004)
9. Hamamci, S.E.: An algorithm for stabilization of fractional-order time delay systems using fractional-order PID controllers. IEEE Trans. Autom. Control **52**, 1964–1969 (2007)
10. Hamamci, S.E., Koksal, M.: Calculation of all stabilizing fractional-order PD controllers for integrating time delay systems. Comput. Math. Appl. **59**, 1621–1629 (2010)
11. Hilfer, R. (ed.): Applications Of Fractional Calculus in Physics. World Scientific, Singapore (2000)
12. Hotzel, R.: Some stability conditions for fractional delay systems. J. Math. Syst. Estimation Control **8**, 1–19 (1998)
13. Hwang, C., Cheng, Y.-C.: A numerical algorithm for stability testing of fractional delay systems. Automatica **42**, 825–831 (2006)
14. Lazarević, M.: Stability and stabilization of fractional order time delay systems. Sci.Tech. Rev. **61**, 31–45 (2011)
15. Marshall, J.E., Górecki, H., Korytowski, A., Walton, K.: Time-Delay Systems: Stability and Performance Criteria with Applications. Ellis Horwood, New York (1992)
16. Nguyen, L.H.V., Bonnet, C.: Stability analysis of fractional neutral time-delay systems with multiple chains of poles asymptotic to same points in the imaginary axis. In: 51st IEEE Conference on Decision and Control, Maui, USA (2012)
17. Partington, J.R., Bonnet, C.: H_∞ and BIBO stabilization of delay systems of neutral type. Syst. Control Lett. **52**, 283–288 (2004)
18. Quadrat, A.: On a generalization of the Youla-Kučera parametrization. Part I: the fractional ideal approach to SISO systems. Syst. Control Lett. **50**, 135–148 (2003)
19. Si-Ammour, A., Djennoune, S., Bettayeb, M.: A sliding mode control for linear fractional systems with input and state delays. Commun. Nonlinear Sci. Numer. Simul. **14**, 2310–2318 (2009)
20. Smith, M.C.: On stabilization and the existence of coprime factorizations. IEEE Trans. Autom. Control **34**, 1005–1007 (1989)
21. Vyhlídal. T., Zítek, P.: QPmR v. 2 Quasipolynomial rootfinder, algorithm and examples. To appear. In: Vyhídal, T., Lafay, J.F., Sipahi, R. (eds.) Advances in Delays and Dynamics. Springer (2013)

Chapter 5
Controller Design for a Class of Delayed and Constrained Systems: Application to Supply Chains

Charifa Moussaoui, Rosa Abbou and Jean Jacques Loiseau

Abstract This chapter aims to investigate the construction of efficient controllers for some input time delay systems, subjected to strict constraints of positivity and saturating limitations, in presence of some exogenous bounded disturbances. The results are presented through the application of supply chains, for which controller design is a challenging issue trading-off between stabilising properties in presence of time lags, and constraints due to the physical limitations and specificities of the plants of the supply chain. We show that the stabilization of such flow systems can be tackled by the stabilization of input time delay systems, using a predictor based feedback approach. This classical control method, which permits to overcome the delays, is enriched by using saturation terms that allow the consideration of the physical constraints of the system resources composing the serial supply chains.

5.1 Introduction

Input time delay systems are common models widely used for dead-time systems representation. Such systems are characterized by the presence of some irreducible time lags, due to, for example, process durations, mass or information transport phenomena and sensor responses. These systems attracted a great deal of attention from both practitioners and researchers, since they are involved in many industrial processes

C. Moussaoui (✉) · R. Abbou · J. J. Loiseau
Institut de Recherche en Communications et Cybernétique de Nantes,
LUNAM Université, UMR 6397. École Centrale de Nantes, 1 rue de la Noë,
BP 92101, 44321 Nantes Cedex 3, France
e-mail: Charifa.Moussaoui@irccyn.ec-nantes.fr

R. Abbou
e-mail: Rosa.Abbou@irccyn.ec-nantes.fr

J. J. Loiseau
e-mail: Jean-Jacques.Loiseau@irccyn.ec-nantes.fr

A. Seuret et al. (eds.), *Low-Complexity Controllers for Time-Delay Systems*,
Advances in Delays and Dynamics 2, DOI: 10.1007/978-3-319-05576-3_5,

and applications. The related control problems are challenging issues which fuel constantly the researchers community. Indeed, in addition to the presence of time-lags, these systems are frequently subject to some physical constraints and additional specificities. Furthermore, they are often subject to some exogenous perturbations effects, which makes the controller design task quite more difficult.

Our concerns focus on the supply lines which are quite representative of such dead-time systems. They consist of a network of interconnected stages, composed of manufacturers, suppliers parts, exchanging goods, financial and information flows, through transportation, warehousing and retailing operations, all in the sake of fulfilling end-customer requests. These operations are time consuming and request important time-lags that can not be neglected or simply approximated. A central issue is to coordinate them over different stages and locations, while providing a convenient service level to end-customers. This task is quietly enhanced when the market demands are unstable or unknown in advance. In addition, the supply line resources are limited by the storage and the production or supplying capacities, which are current bottlenecks for the system. Indeed, the storage devices, say the inventories, are finite resources that can be subject to congestion problems which lead to important goods losses, while the production units are subject to saturation phenomena due to actuator limitations. These practical problems are commonly performed in other systems, where the dynamics is governed by constrained flow exchanges, such as the communication networks and some others load-balancing systems, where the information flows and the buffers in the network nodes, can be perceived, respectively, as the good flows and the inventories of a supply line.

In this chapter, supply line control issue is formulated as a general constrained control problem, for input time delay systems with positivity constraints and saturated resources, subject to unknown but bounded disturbances. To handle this problem, we make use of predictive-based control techniques, which efficiency in compensating the input time delay is well-established and widely described in the literature. Using pole assignment principle and model reduction [1, 11, 13], we propose a saturated and constrained control law, which allows the controller to handle the system constraints and to meet its specifications. The proposed methodology for designing such controller consists in defining an invariant set for the system trajectories, such that the bounded input bounded output (BIBO) stability property is ensured, and for which the system constraints are meet.

After introducing the problem, the chapter is organized as follows. In Sect. 5.2, the dynamical model of each stage of the considered supply line is proposed, and a quick review of the literature concerning this topic is presented. Section 5.3 is devoted to the controller structure, where some backgrounds about saturated commands and the predictive control methods for input time delay systems are presented. In Sect. 5.4, the controller design issues are addressed and these results are extended to the case of a multi-level supply line in the Sect. 5.5, and illustrated through an numerical example in Sect. 5.6. Finally, a discussion about the obtained results and further investigations conclude this chapter.

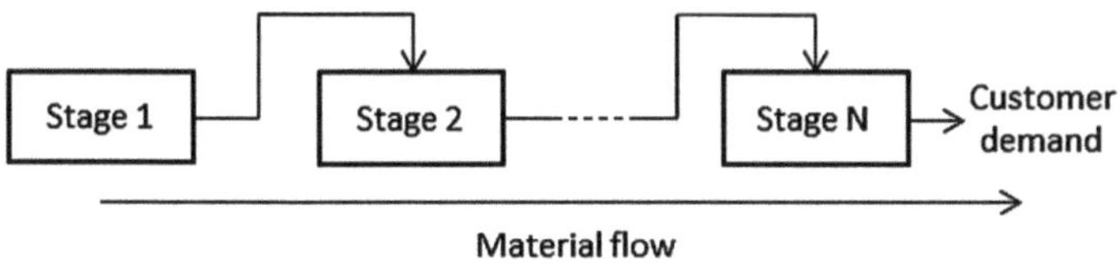

Fig. 5.1 Multi stages supply chain scheme

5.2 Problem Statement: Inventory and Production Control

A supply line is a series of stages or levels, which represent manufacturers, suppliers, transporters and other parts that are involved in the supplying process, in which the goods flow linearly to reach end-customers, as depicted in Fig. 5.1.

Supply chain control consists of defining appropriate ordering policies that regulate the production and the supply rates in the different stages of the supply line, so that each stage is able to meet the inventories requirements, and to provide a good service level for the incoming demands. In this field, different frameworks where proposed during the past decades, based on optimization procedures using programming techniques, empirical experiences and control theory methods. Our concern focused on the use of the control theory methods, which provide an analytical and formal framework and allow a structural approach to handle the supply chain issues. Indeed, since the pioneering works of Simon [18], who was the first to use Laplace transform to analyze a supply line dynamics, numerous investigations followed, such that [4, 6, 15, 22, 24, 28], in which the supply chain was modelled using block diagrams and controlled through feedback structures. These investigations lead to the well-known Automatic Pipeline Inventory and Order Based Production Control System (API-OBPCS) models and their variations [10]. They permit to understand the complex interactions that govern supply chains dynamics, identifying the critical agents that impact the inventories stability, such that the delays. The authors highlighted the importance of the Work In Process (WIP), which is the amount of goods ordered in the pipeline but not yet received due to the delay. They also reveal its central role in damping the variance of the demand amplification among the supply chain stages, which is known as the bullwhip or Forrester effect [6].

The advances in the time delay systems control [14, 21], allow further insights into the delayed differential equations describing , in particular, the inventory dynamics [3, 19, 27], and notable works like [20] permit great extensions considering multiple delays. Nevertheless, the aforementioned works did not take into account the positivity and the capacity constraints of the supply chain resources. Actually, both inventory levels and replenishment orders are constraint free, and are allowed to get some negative values or excessive huge ones, which does not correspond to real plant capacities and thus creates a major gap between theoretical attempts and practical results. For such issues, simulation based analyses are the mostly used, such as in [2, 5, 7, 8, 17, 29], where the impact of constrained production capacity is studied. In [25, 26] an analytical investigation is presented for the forbidden-return case, which corresponds to the constraint of non-negativity on the replenishment orders only. These studies pointed out that considering capacity constrained on the supplying devices, removes the linearity assumption of the model and hence complex dynamics behaviours are revealed. To the best of our knowledge, no work in this field considers capacity and positivity constraints, on both the supplying devices and the inventories, taking into account the pure delays present in these systems dynamics. This is what this work contributes to. We consider a multi-stage supply line. In a first attempt, our analysis will be held for a single stage, the general case will be presented in Sect. 5.5.

Each stage of the supply line represents an elementary system composed of a supplying unit and a storage one. The term "supplying" is used for the material supplying acquisition operations, which can be production, transport or retailing process. The supplying units are characterized by a delay θ which corresponds to the time needed to complete the supplying task, and by a supplying order rate denoted $u(t)$, which is limited by a maximum supplying capacity denoted U_{max}. The storage units are namely the inventories. Each elementary stage of the supply chain has an inventory with a maximum storage capacity denoted Y_{max}. In this work, the customer demands are unknown in advance but assumed to be upper bounded by an amount denoted d_{max}. The generic model for the inventory level dynamics is then described by the following first order delayed equation.

$$\dot{y}(t) = \begin{cases} u(t-\theta) - d(t) & \text{for } t \geq \theta \\ \phi(t) - d(t) & \text{for } 0 \leq t < \theta \, , \end{cases} \tag{5.1}$$

where, $y(t)$ is the inventory level and $d(t)$ the incoming demand rate of each level. The function $\phi(t)$ describes the initial state of the system such that Eq. (5.1) describes the initial dynamics of the inventory for $0 \leq t < \theta$.

As already mentioned, supplying units and inventories as well, are limited resources, which can take non-negative values only. These constraints are formulated as follows. For inventory level

$$y(t) \in [0, Y_{max}] \, , \text{ for } t \geq 0 \, , \tag{5.2}$$

and for the supplying rate

$$u(t) \in [0, U_{max}] \, , \text{ for } t \geq 0 \, . \tag{5.3}$$

Then, the working assumption on the consumer demand is formulated such as

$$d(t) \in [0, d_{max}] \, , \text{ for } t \geq 0 \, . \tag{5.4}$$

The controller design task consists of defining a controller which will stabilize the delayed system (5.1) while ensuring the fulfilment of the constraints (5.2) and (5.3), for every bounded disturbance verifying the assumption (5.4).

5.3 System Control Structure

A local management strategy is used for each elementary level, which aims at meeting its local specifications. In this work, we consider that all stages of the considered supply line are applying the same ordering rules, which consists of fulfilling on line the consumer demands, and replenishing the inventory to a referential level denoted y_c. The strategy to define on line the control law, that is the supplying rate at each level of the supply chain, is presented in the following section.

5.3.1 *Order Rates and Control Structure*

The order rate $u(t)$ at a given level represents the command of the delayed system given by Eq. (5.1). Regarding to the system constraints, and the nature of the system, the control law we propose to apply is *a saturated command based on a feedback predictor structure* such that

$$u(t) = \underset{[0, U_{max}]}{sat} \left[K(y_c - z(t)) \right], \text{ for } t \geq 0, \tag{5.5}$$

where y_c is the reference signal of the system, which corresponds to the reference level for the inventory. K is the controller gain which is used to adjust the order rates placed in each level, and $z(t)$ is the prediction of the future state of the system, that corresponds to the inventory level at $t + \theta$, as it is shown in the sequel.

Saturated commands are commonly used for systems with saturating actuators, and permit to take into account theirs specific limitations. It was shown to be more efficient and realistic than a linear constraint control [9, 23]. On the other hand, the use of a saturated controller introduces non-linearities in the closed-loop scheme of the system, due to the *sat* function defined as

$$\underset{[a,b]}{sat} [f(t)] = \begin{cases} b & \text{if } f(t) > b\,, \\ f(t) & \text{if } a \leq f(t) \leq b\,, \\ a & \text{if } f(t) < a\,. \end{cases}$$

For such non-linear systems, stability conditions can be obtained by computing invariant sets in which the system trajectory remains, and in which the saturation constraints are met, as it is shown in the sequel. The feedback predictor part of the command, is used to handle the delays and the stability properties of the infinite-dimensional system, by allowing the assignment of the closed-loop system poles, in a finite number of locations in the complex plan [11, 13]. Also known as model reduction or Artstein reduction [1], the basic idea of state prediction is to compensate the time delay θ by generating a control law that enables one to directly use the corresponding delay-free system, thanks to the prediction defined by

$$z(t) = \begin{cases} y(t) + \int_{t-\theta}^{t} u(\tau)d\tau & \text{for } t \geq \theta\,, \\ y(t) + \int_{t}^{\theta} \phi(\tau)d\tau + \int_{0}^{t} u(\tau)d\tau & \text{for } t < \theta\,, \end{cases} \tag{5.6}$$

which can be rewritten, using expression (5.1) as

$$z(t) = y(t+\theta) + \int_{t}^{t+\theta} d(\tau)d\tau\ , \ \text{for } t \geq 0. \tag{5.7}$$

Indeed, by time derivation of this Eq. (5.7), one can see that the resulting system

$$\dot{z}(t) = u(t) - d(t)\ , \text{for } t \geq 0\,, \tag{5.8}$$

is delay-free. The system (5.8) is the reduced model of the system (5.1)–(5.5). Artstein [1] demonstrated that the control low $u(t)$ is admissible for the closed loop system (5.1)–(5.5) if and only if it is admissible for the system (5.8)–(5.5), and that the two systems have the same dynamics properties. Our approach is then based on the use of the reduced system (5.8) to design the controller such that the system constraints and requirements (5.2) and (5.3) will be fully met, as shown in Sect. 5.3.2.

5.3.2 The Closed-Loop System Dynamics

The dynamics of the closed-loop system (5.8)–(5.5) is given by the following expression.

$$\dot{z}(t) = \underset{[0, U_{max}]}{sat} \left[K(y_c - z(t)\right] - d(t)\ , \ \text{for } t \geq 0\,. \tag{5.9}$$

The stability analysis of this system is performed by computing an invariant interval for the trajectories of system (5.9), in which the system constraints are met, and the BIBO stability property of the system is warranted. In this sake, the system constraints (5.2) and (5.3) are reformulated in terms of the new state variable $z(t)$ as follows.
Using the expression (5.7), one can see that

$$y(t+\theta) = z(t) - \int_{t}^{t+\theta} d(\tau)d\tau\ , \ t \geq 0. \tag{5.10}$$

The constraint (5.2) is verified if both $z(t)$ and the term $\int_t^{t+\theta} d(\tau)d\tau$ are bounded, so that $y(t+\theta) \in [0,\ Y_{max}]$. Provided that condition (5.4) is satisfied, it is seen that $\int_t^{t+\theta} d(\tau)d\tau \in [0, \theta\, d_{max}]\ , \forall t \geq 0$. Thus, $z(t)$ should be limited by a lower and an upper bound, z_{min} and z_{max} respectively, which verifies the relation (5.10), such that

$$0 \leq z_{min} - \theta d_{max} \text{ for } y(t+\theta) = 0\,, \quad \text{and} \quad z_{max} \leq Y_{max} \text{ for } y(t+\theta) = Y_{max}\,.$$

Then, the original delayed system verifies $y(t) \in [0, Y_{max}]$ for all $t \geq 0$, if and only if the condition

$$z(t) \in [\theta d_{max}, Y_{max}]\,, \tag{5.11}$$

with $\theta d_{max} < Y_{max}$, is verified for the reduced system. The control problem established in Sect. 5.2, is reformulated in terms of founding the controller parameters, which permits to the closed loop system (5.8)–(5.7), to verify the constraints (5.11) under disturbance effects of $d(t)$. The results are given in Sect. 5.4.

5.3.3 Admissible Initial Conditions

Non-zero initial conditions does not affect the control structure and the system constraints. Indeed, as shown by Eq. (5.1) for $t \in [0, \theta]$, the inventory level evolution depends on the function $\phi(t)$ and the demand only. Because of the delay, the effects of the command $u(t)$ on the system dynamics are not visible before $t = \theta$. Then, checking whether the system constraints are met or not on the time interval $[0, \theta]$ yields a set of admissible initial conditions, for which the constraint conditions are verified. This set is characterized as follows. Using Eq. (5.1), for $0 \leq t < \theta$, the inventory level is given by

$$y(t) = y_0 + \int_0^t \phi(\tau)\,d\tau - \int_0^t d(\tau)\,d\tau\,,$$

where y_0 is the initial inventory level at time $t = 0$, and the amount $\int_0^\theta \phi(\tau)d\tau$ represents the initial WIP in the pipeline that is denoted wip_0. It is seen that the term $\int_0^t d(\tau)d\tau$ belonging to the interval $[0, \theta d_{max}]$, $y(t)$ verifies $y(t) \in [0, Y_{max}]$ for $t \in [0, \theta]$ if and only if the initial conditions are such that

$$\theta\, d_{max} \leq y_0 + wip_0 \leq Y_{max}.$$

5.4 Controller Designing Issues

The controller design consists in determining suitable gain K and inventory reference level y_c for each elementary stage of the supply chain, such that the system constraints and specifications are fully met. Two main issues are to be considered. First, for given systems parameters, namely the maximum capacities U_{max} and Y_{max}, is it possible to find a controller which will fully meet the constraints and the system requirements. Then, if such a controller is indeed feasible, the second issue is about the choice of the command parameters K and y_c under the system constraints and specifications. This is the parameterization phase. In this section, both issues are treated through the dynamics properties analysis of the system, such that the exact solution of the Eq. (5.9) is not required. Our proposal is to determine some necessary and sufficient conditions on the controller parameters, to impose the invariance property of the interval (5.11), so that the BIBO stability of the system and the constraints are all satisfied. These conditions are given through the following Theorem 1. A corollary is then formulated, which gives further results concerning the closed-loop system dynamics under Theorem 1 assumptions.

Theorem 1 *Being given a system of the form (5.1), there exists a command of the form (5.5), for which the system is stable and the constraints (5.2) and (5.3) are fulfilled, for any* $d(t) \in [0, d_{max}]$ *if and only if the following conditions hold true*

$$\theta d_{max} < Y_{max} , \tag{5.12}$$

and

$$d_{max} \leq U_{max} . \tag{5.13}$$

In addition, if the conditions (5.12) and (5.13) are met, the constraints (5.2) and (5.3) are satisfied under the control law (5.5) if an only if the controller parameters are such that:

$$\theta d_{max} + \frac{d_{max}}{K} \leq y_c \leq Y_{max}. \tag{5.14}$$

Proof As shown in Sect. 5.3.2, the controller of the reduced system (5.9) should be designed such that constraint (5.11) is fulfilled. The existence of the controller is then linked with the non-empty property of interval $[\theta\, d_{max}, Y_{max}]$, which is true only when $\theta\, d_{max} < Y_{max}$. This later shows the necessity of condition (5.12), its sufficiency being obvious.

Conditions (5.13) and (5.14) come from the fact that, verifying constraint (5.11) at any time $t \geq 0$ implies that, the closed interval $[z_{min}, z_{max}]$ is invariant for the system trajectories. Formally, this property is warranted if and only if the following implications are true, for all $t \geq 0$

$$z(t) = z_{min} \Rightarrow \dot{z}(t) \geq 0\,, \quad \text{and} \quad z(t) = z_{max} \Rightarrow \dot{z}(t) \leq 0\,.$$

Using expression of $\dot{z}(t)$ given by (5.9), and provided that (5.4) is true, these inequalities are rewritten respectively

$$\underset{[0,U_{max}]}{sat} \left[K(y_c - z_{min})\right] \geq d_{max}\,, \tag{5.15}$$

and

$$\underset{[0,U_{max}]}{sat} \left[K(y_c - z_{max})\right] \leq 0\,. \tag{5.16}$$

Using the *sat* function definition given in Sect. 5.3.1, one can see that the inequality (5.15) is solvable if and only if $U_{max} \geq d_{max}$, that shows the sufficiency and necessity of condition (5.13), and thus y_c is such as $z_{min} + d_{max}/K \leq y_c$, which, together with condition (5.12) and the equality $z_{min} = \theta d_{max}$, establishes the sufficiency of the left part of the condition (5.14) of Theorem 1. Its necessity comes form the fact that for $y_c < z_{min} + d_{max}/K$, inequality (5.15) has no solution.

The same analysis is applied for inequality (5.16). This latter is solvable if and only if

$$y_c \leq z_{max}, \tag{5.17}$$

which, together with equality $z_{max} = Y_{max}$ shows that $y_c \leq Y_{max}$. This establishes the sufficiency and the necessity of the right member of condition (5.14) of Theorem 1, and completes the proof. □

Under the conditions of Theorem 1, the analysis of the closed-loop system dynamics shows that the system constraints and specifications are truly met. Any satisfactory controller actually permits to fill more restrictive constraints on the system variables. We describe these restrictions in the following corollary.

Corollary 1 *Being given a system of the form (5.1), with a control law of the form (5.5) and suitable initial conditions, such that the conditions (5.12), (5.13) and (5.14) are verified, then the following holds true*

$$y(t) \in \left[y_c - \frac{d_{max}}{K} - \theta\, d_{max} \,, y_c \right], \tag{5.18}$$

and

$$u(t) \in [0 \,, d_{max}] \,, \tag{5.19}$$

for all $t \geq 0$ *and* $d(t) \in [0, d_{max}]$.

Proof From expression (5.9), one can observes that, under Theorem 1 assumptions, the following implications are true for all $t \geq 0$,

$$z(t) \geq y_c \Rightarrow \dot{z}(t) \leq 0 \quad \text{and} \quad z(t) \leq y_c - \frac{d_{max}}{K} \Rightarrow \dot{z}(t) \geq 0.$$

These implications show that the effective interval of variation of $z(t)$ is such that

$$z(t) \in \left[y_c - \frac{d_{max}}{K} \,, y_c \right], \tag{5.20}$$

which represents the smallest invariant interval for the system (5.9). Indeed, under Theorem 1 assumptions, it is seen that the interval given in (5.20) is included in the interval given by (5.11). Thus, using expressions (5.10) and (5.5), one can compute the effective interval of $y(t)$ and $u(t)$ variations which are given by expressions (5.18) and (5.19) respectively. □

5.5 Generalization for N-Stages Supply Chain

In this section, we propose a generalization of the results presented above for the multi-stages supply line, composed of N elementary stages, as presented in Sect. 5.2. Each level is now labelled with a subscript i, with $i = 1, .., N$. In such serially linked structure, each stage i has one supplier $i-1$, and is supposed to support the incoming demand $d_i(t)$ of the following stage $i+1$ such that $d_i(t) = u_{i+1}(t)$ for $i = 1, .., N-1$. The last stage N of the supply chain is usually the final retailer which faces the end consumer demand denoted $d_c(t)$. The inventory dynamics of each stage is given by the following equation.

$$\dot{y}_i(t) = \begin{cases} u_i(t-\theta_i) - d_i(t) & \text{for } t \geq \theta_i, \\ \phi_i(t) - d_i(t) & \text{for } 0 \leq t < \theta_i, \end{cases} \tag{5.21}$$

where, $y_i(t)$ is the inventory, $u_i(t)$ is the acquisition rate with delay θ_i, and $d_i(t)$ the incoming demand, with $i = c$ for the customer demand rates.

The present part aims to define the controller parameters of each stage, such as the end consumer demand $d_c(t)$ will be satisfied, and taking into account the local constraints of each single level, as seen in Sect. 5.4, and the additional constraints due to the serial structure as well. The constraints (5.2) and (5.3) are generalized as follows. For $i = 1, .., N$, the inventory levels are such that

$$y_i(t) \in \left[0, Y_{max_i}\right], \text{ for } t \geq 0, \tag{5.22}$$

and the acquisition rates verify

$$u_i(t) \in \left[0, U_{max_i}\right], \text{ for } t \geq 0. \tag{5.23}$$

The additional constraints arising from the network structure are about the incoming demand of each level, where $d_i(t) = u_{i+1}(t)$ for $i = 1, .., N-1$, such that

$$d_i(t) \in \left[0, U_{max_{i+1}}\right]. \tag{5.24}$$

For the retailer stage $i = N$, the incoming demand is the end customer demand $d_c(t)$. It verifies the same assumption (5.4) namely

$$d_c(t) \in [0, d_{max}]. \tag{5.25}$$

The same control law as the one presented in Sect. 3.2, is used in each stage. It is of the form

$$u_i(t) = \underset{[0, U_{max_i}]}{sat} \left[K_i(y_{c_i} - z_i(t))\right] \text{ for } t \geq \theta_i, \tag{5.26}$$

with y_{c_i} is the reference level for the inventory y_i, K_i is the controller gain which is used to adjust the order rates placed in level i, and $z_i(t)$ is the prediction of the future state of the system, defined as follows

$$z_i(t) = \begin{cases} y_i(t) + \int_{t-\theta_i}^{t} u_i(\tau)\, d\tau & \text{for } t \geq \theta_i, \\ y_i(t) + \int_{t}^{\theta_i} \phi_i(\tau)\, d\tau + \int_{0}^{t} u_i(\tau)\, d\tau & \text{for } t < \theta_i. \end{cases} \quad (5.27)$$

The controller design issue in each stage is addressed as shown in Sects. 5.3 and 5.4. Using the same arguments basing on the analysis of the reduced model obtained for each stage, Theorems 1 is now extended to the N-stages supply line.

Theorem 2 *Being given a supply chain of the form (5.21), there exists a command of the form (5.26), for which the system is stable, and fulfills the constraints (5.22), (5.23) and (5.24), for any* $d_c(t) \in [0, d_{max}]$ *if and only if the following conditions hold true.*

$$\theta_i\, d_{max} < Y_{max_i}, \quad (5.28)$$

and

$$d_{max} \leq U_{max_i}, \quad (5.29)$$

for all $t \geq 0$ *and* $i = 1, .., N$. *In addition, if conditions (5.28) and (5.29) are verified, the constraints (5.23), (5.24) and (5.25) are met for any* $d_c(t) \in [0, d_{max}]$, *if and only if the controller parameters are such that:*

$$\theta_i\, d_{max} + \frac{d_{max}}{K_i} \leq y_{c_i} \leq Y_{max_i}, \quad (5.30)$$

for all $t \geq 0$ *and* $i = 1, .., N$.

Proof For the N-stages supply line serially linked, the whole supply line dynamics is driven by the end-costumer demand $d_c(t)$. Applying Theorem 1 for the last stage N of the line, Theorem 2 shows that the orders $u_N(t)$ vary in the interval $[0, d_{max}]$. Then the linking relation between the supply line stages, where $d_{N-1}(t) = u_N(t)$ shows that actually $d_{N-1}(t) \in [0, d_{max}]$. Thus, by recursion, it is seen that $d_i(t) \in [0, d_{max}]$ for all $i = 1, .., N$. Using this result, the demonstration of Theorem 2 is directly derived from the proof of Theorem 1. □

5.6 Simulation Example and Discussions

The application example presented in this section, aims at highlighting the efficiency of the distributed control scheme proposed to eliminate the bullwhip effect in a supply chain, and to illustrate the importance of taking into account the

Table 5.1 Simulation parameters for the constrained three-stages supply chain

Stage	K_i	θ_i	U_{max_i}	Y_{max_i}	y_{ci}	$y_{0i} + wip_{0i}$
$i = p$	0.6	6	350	1900	1900	1900
$i = w$	0.6	3	300	1200	1200	1200
$i = r$	0.6	2	250	900	900	900

positivity and capacity constraints. For this sake, we consider a three-stages supply chain, commonly used in the literature [5, 8], consisting of a production plant, a wholesale stage, and a distribution centre. Unlike the aforementioned works, where capacity constraints are assumed for the order rates only, we consider both inventories limitations, and positivity constraints. The subscripts p, w, and r are used to label the production, the warehouse and the retailer stages respectively. The maximum capacities U_{max_i} and Y_{max_i} and some admissible initial conditions are given in Table 5.1.

The initial conditions are chosen such that all transitory dynamics are avoided, according to the rule defined in Sect. 3.3. Controller parameters K_i and y_{ci} are calculated according to Theorem 2, and are also sorted in Table 5.1. In order to illustrate the inventory dynamics, the customer demand used for this simulation is a square function starting at $t = 15$ weeks and ending at $t = 45$ weeks, with an amplitude $d_{max} = 240$, unlike the aforementioned works where only step function demands where considered. The results of the simulation are depicted on Fig. 5.2, where the order rates of each stage and the inventory levels are represented.

Figure 5.2a shows that the order rates placed in each of the three stages, follows closely the demand, causing no amplification through the upstream levels as it is expected by the relation (5.24). Then, Fig. 5.2b shows that the inventory levels remain non-negative, and are re-completed to their reference levels when the demand is null, as it was specified by the ordering policy presented in Sect. 5.1. The outcomes of this simulation study, join the former authors conclusions [2, 5, 8, 29] which stand that capacity constraint of the ordering rates, does not necessarily impact the customer service level, which corresponds to the demand satisfaction. The saturating constraint impacts the dynamics too, such that the order rates being limited, the supplying process completion takes more time, but since the condition (5.29) is verified, the demand is always fulfilled. It is also recognized that such constraints provide an effective improvement in reducing the demand amplification, within the multi-echelon system. Indeed, we showed that a good handling of the delays, via an appropriate control law, permits to definitely overcome the Forrester effect. This result is also pointed out by the former work of [16] and [18], where it is established that the smoothest system responses are obtained when the same care is given to the inventory discrepancy and the WIP. The formal explanation of this empiric result comes from the input time delay system control, as seen in Sect. 5.3, where the efficient delay compensation via the predictor feedback imposes the same coefficient K for both the inventory discrepancy and the distributed delay of the predictor which

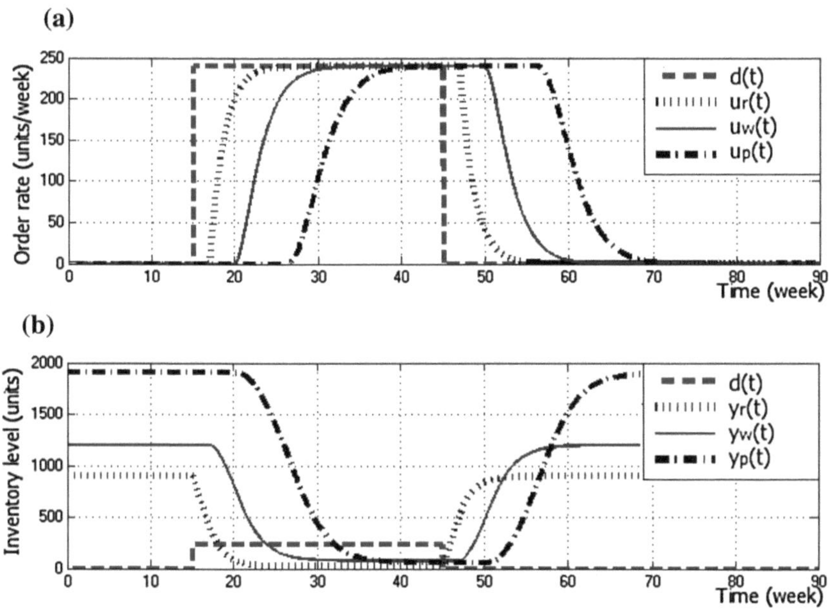

Fig. 5.2 Simulation results for the constrained three-stages supply line. **a** Dynamics of the order rates, **b** Dynamics of the inventory levels

is the WIP term. Assuming an unknown bounded demand as a working assumption, allows us to maintain this results for every bounded demand signal, no mater if it is a step function shaped or not.

5.7 Conclusion and Perspectives

In this chapter, the controller design problem for serially-linked supply chains, with constrained orders and inventories, and unknown customer demands variations, has been investigated. The problem is stated in terms of controlled input time delay system, with positivity and saturations constraints, subject to bounded disturbances. A saturated feedback predictor controller was introduced to handle both the delayed dynamics and the constraints, where the controller encompasses a distributed delay expressed by the integral term in the prediction. This distributed term corresponds to the WIP amount which the use in inventory regulation is quite classical for damping the bullwhip effect [10, 16, 18]. It is important to notice that the WIP is actually measurable. Thus, the controller proposed in this work is of low complexity, since it corresponds to a static feedback on measurable variables. The main advantage of this work is that practical constraints of positivity and capacity of both orders and inventories are taken into account, that enhanced the accuracy of the results.

In addition, the controller proposed eliminates totally the Forrester effect, in case where the delays are properly known. Robustness analysis of the results in case of delay misestimations, and the consideration of variable delays are advised of forthcoming works.

References

1. Artstein, Z.: Linear systems with delayed controls: a reduction. IEEE Trans. Autom. Control **27**(4), 869–879 (1982)
2. Cannella, S., Ciancimino, E.: Capacity constrained supply chains: a simulation study. Int. J. Simul. Process Model. **4**(2), 139–147 (2008)
3. Delice, I.I., Sipahi, R.: Inventory dynamics models of supply chains with delays; system-level connection and stability. In: Loiseau, J.J., Michiels, W., Niculescu, S.-I., Sipahi, R. (eds.) Topics in Time Delay Systems, of Lecture Notes in Control and Information Sciences, vol. 38, pp. 8349–8358. Springer, Berlin Heidelberg (2009)
4. Edghill, J.S., Towill, D.R.: The use of systems dynamics in manufacturing systems. Trans. Inst. Meas. Control **11**(4), 208–216 (1989). doi:10.1177/014233128901100406
5. Evans, G.N., Naim, M.M.: The dynamics of capacity constrained supply chains. In: Proceedings of International System Dynamics Conference, pp. 28–35. Stirling, Scotland (2008)
6. Forrester, J.W.: Industrial Dynamics. MIT press, Cambridge (1961)
7. Gavirneni, S., Kapucinski, R., Tayur, S.: Value of information in capacitated supply chains. Manage. Sci. **45**(1), 16–24 (1999)
8. Helo, P.T.: Dynamic modelling of surge effect and capacity limitation in supply chains. Int. J. Prod. Res. **38**(17), 4521–4533 (2000)
9. Hu, T.L.Z.: Control Systems with Actuator Saturation: Analysis Ddesign. Birkhäuser, Boston (2001)
10. John, S., Naim, M.M., Towill, D.R.: Dynamic analysis of a WIP compensated decision support system. Int. J. Manage. Syst. Des. **1**(4), 283–297 (1994)
11. Kwon, W., Pearson, A.: Feedback stabilization of linear systems with delayed control. IEEE Trans. Autom. Control **25**(2), 266–269 (1980)
12. Manitius, A., Olbrot, A.: Finite spectrum assignment problems for systems with delays. IEEE Trans. Autom. Control **24**, 541–553 (1979)
13. Mirkin, L., Raskin, N.: Every stabilizing dead-time controller has an observer-predictor-based structure. Automatica **39**(10), 1747–1754 (2003)
14. Richard, J.P.: Time-delay systems: an overview of some recent advances and open problems. Automatica **39**(10), 1667–1694 (2003)
15. Riddalls, C.E., Bennett, S., Tipi, N.S.: Modeling the dynamics of supply chains. Int. J. Syst. Sci. **31**, 969–976 (2000)
16. Riddalls, C.E., Bennett, S.: The stability of supply chains. Int. J. Prod. Res. **40**(2), 459–475 (2002)
17. Shukla, V., Naim, M.M.: The impact of capacity constraints on supply chain dynamics. In: International Conference on Computers and Industrial Engineering, pp. 925–930. CIE (2009)
18. Simon, H.A.: On the application of servomechanism theory in the study of production control. Econometrica **20**, 247–268 (1952)
19. Sipahi, R., Delice, I.I.: Supply Network Dynamics and Delays; Performance, Synchronization, Stability. Mechanical and Industrial Engineering Faculty Publications, Seoul (2008). http://works.bepress.com/rsipahi/13
20. Sipahi, R., Delice, I.I.: Stability of inventory dynamics in supply chains with three delays. Int. J. Prod. Econ. **123**, 107–117 (2010)

21. Sipahi, R., Niculescu, S.-I., Abdallah, C.T., Michiels, W., Gu, K.: Stability and stabilization of systems with time delay, limitations and opportunities. IEEE Control Syst. Mag. **31**(1), 38–65 (2011)
22. Sterman, J.D.: Modelling managerial behaviour misinterpretations of feedback in a dynamic decision-making experiment. Manage. Sci. **35**(3), 321–339 (1989)
23. Tarbouriech, S., Garcia, G., Da Silva, J., Queinnec, I.: Stability and Stabilization of Linear Systems with Saturating Actuators. Springer, London (2011)
24. Towill, D.R.: Dynamic analysis of an inventory and order based production control system. Int. J. Prod. Res. **20**(6), 671–687 (1982)
25. Wang, X., Disney, S.M., Wang, J.: Exploring the oscillatory dynamics of a forbidden returns inventory system. Int. J. Prod. Econ. **147**, 3–12 (2012). (ISSN 0925–5273) http://dx.doi.org/10.1016/j.ijpe.2012.08.013
26. Wang, X., Disney, S.M., Wang, J.: Stability analysis of constrained inventory systems with transportation delay. Eur. J. Oper. Res. **223**(1), 86–95 (2012)
27. Warburton, R.D.H.: An exact analytical solution to the production inventory control problem. Int. J. Prod. Econ. **92**, 81–96 (2004)
28. Warburton, R.D.H., Disney, S.M., Towill, D.R., Hodgson, J.P.E.: Further insights into 'the stability of supply chains'. Int. J. Prod. Res. **42**(3), 639–648 (2004)
29. Wikner, J., Naim, M.M., Rudberg, M.: Exploiting the order book for mass customized manufacturing control systems with capacity limitation. IEEE Trans. Eng. Manage. **54**(1), 145–155 (2007)

Chapter 6
Delay Effects in Visual Tracking Problems for an Optronic Sighting System

Alban Quadrat and Arnaud Quadrat

Abstract In this chapter, we study the delay effects in visual tracking problems for an optronic sighting system. We first describe the physical model and then give a simplified version defined by an integrator and a time-delay. We then state the visual tracking problems that are considered. To solve these problems, we first have to study the stabilization problem for the system defined above. Since this problem is a particular case of the general problem of parametrizing all the stabilizing controllers of a stable perturbation of a (infinite-dimensional) stabilizable plant, this problem is studied in its generality. Within the fractional representation approach to synthesis problems, we give an elementary proof for the existence of a general parametrization of all the stabilizing controllers of a stabilizable plant which does not necessarily admit doubly coprime factorizations. Only the knowledge of a (finite-dimensional) stabilizing controller is required. If the plant admits doubly coprime factorizations, then this parametrization yields the Youla-Kučera parametrization. Finally, using the above results, we study the tracking problems and show numerical simulations in which our results are compared with a PID and a H_∞-controller.

6.1 Automatic Visual Tracker

In practice, a gyrostabilized optronic payload orients its line of sight in the space [3, 5]. In this chapter, we shall only consider a simplified version, namely the plane case. A visual tracker is a combination of the following four elements [2, 4]:

A. Quadrat (✉)
INRIA Saclay—Île-de-France, Projet DISCO, L2S, Supélec, 3 rue Joliot Curie, 91192 Gif-sur-Yvette cedex, France
e-mail: alban.quadrat@inria.fr

A. Quadrat
SAGEM DS—Etablissement de Massy, 100 avenue de Paris, 91344 Massy Cedex, France
e-mail: arnaud.quadrat@sagem.com

A. Seuret et al. (eds.), *Low-Complexity Controllers for Time-Delay Systems*, Advances in Delays and Dynamics 2, DOI: 10.1007/978-3-319-05576-3_6,

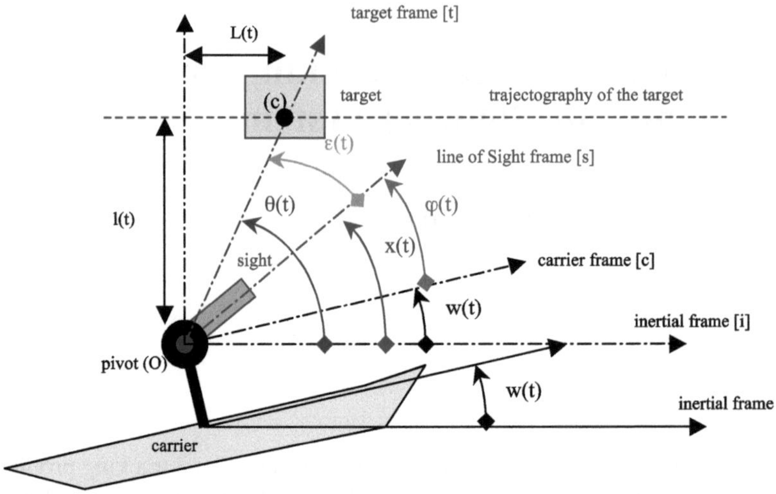

Fig. 6.1 Inertially stabilized boat camera platform

- A gyrostabilized optronic payload is a torque motorized speed controlled platform using the gyrometer speed.
- The optics consists of a digital video camera or a thermal imager set on the platform.
- Using image processing, an automatic image tracker detects the target and returns its coordinates in the frame images with a certain delay.
- A tracker controls the inertial speed of the inertially stabilized imager.

Let $[i]$ be the inertial frame with origin (O). See Fig. 6.1.

To each body, we attach a frame at (O): $[c]$ is the frame of the carrier, $[s]$ is the frame of the line of sight of the video camera and $[t]$ is the frame attached to the *target* located at the point (C). The line of sight can rotate from the carrier $[c]$ thanks to a motorized pivot linkage at (O). Let us now introduce the different angles:

- x is the angle defined by the line of sight of the camera in the inertial frame $[i]$.
- θ is the polar coordinate of the target in the inertial frame $[i]$.
- The angle between $[t]$ and $[s]$ is given by the image obtained by the camera:

$$\varepsilon := \theta - x. \tag{6.1}$$

The angle ε between the target and the line of sight is not directly accessible. To measure it, one can use an image processing device called the *image tracker*. See Fig. 6.2.

The coordinates of the target are determined by means of digit image correlation techniques or *centroid detection* methods. The position of the target can be characterized in terms of a shift of N pixels from the center of the image. At the focal distance of the camera D_f and the size of the optical sensor D, we have $\varepsilon \simeq \tan \varepsilon = \frac{D\,N}{D_f\,N_{\max}}$. The image processing introduces delays and two constraints:

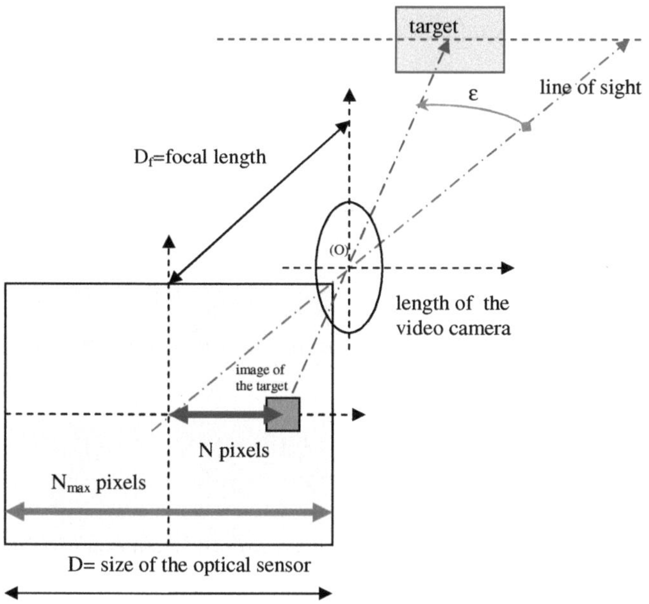

Fig. 6.2 Image tracker

- The image tracker yields a time-delay $T \in [0.04\,\text{s}, 0.2\,\text{s}]$ and a distributed delay $\tau_i \in [1\,\text{ms}, 40\,\text{ms}]$ (which can be neglected), and is corrupted by a noise u_1:

$$\widehat{e}_1 = \widehat{u}_1 - G\widehat{\varepsilon}, \quad G := e^{-Ts}\,\frac{1 - e^{-\tau_i s}}{\tau_i\, s} \simeq e^{-Ts}. \tag{6.2}$$

- The image of the target must stay within the image of the video camera, i.e., $|\varepsilon(t)| \leq C_1$ for all t, and the signal processing imposes $|\dot{\varepsilon}(t)| \leq C_2$ for all t.

The movement of the target is unknown. Its cartesian coordinates are $(L,\ l)$ in the inertial frame $[i]$. See Fig. 6.1. The trigonometric relation between the polar coordinates and the cartesian ones yield $\theta = \arctan(l/L)$. If we set $\theta_0^{(n)} := \theta^{(n)}(0) \in \mathbb{R}$, where $\theta^{(n)}$ denotes the nth derivative of θ with respect to t, then the following three scenarios are admissible:

- **Scenario 1:** Constant position: $\theta = \theta_0$.
- **Scenario 2:** Constant angular speed: $\theta = {\theta_0}^{(1)}\, t + \theta_0$.
- **Scenario 3:** Constant angular acceleration: $\theta = {\theta_0}^{(2)}\, t^2 + {\theta_0}^{(1)}\, t + \theta_0$.

A gyrometer observes the speed $\dot{x}$ of the line of sight. The gyrostabilized platform is modeled by an inner speed loop used by the outer video tracking loop. The transfer function from the reference speed of the inner loop y_1 to the real sight speed $\dot{x}$ can be written as $s\widehat{x} = F\widehat{y}_1$, where F is a low-pass filter and $\lim_{s\to 0} F(s) = 1$. For more details, [9]. We consider a controller $\widehat{y}_1 = C'\widehat{e}_1$ from the output signal $\widehat{e}_1$ of the

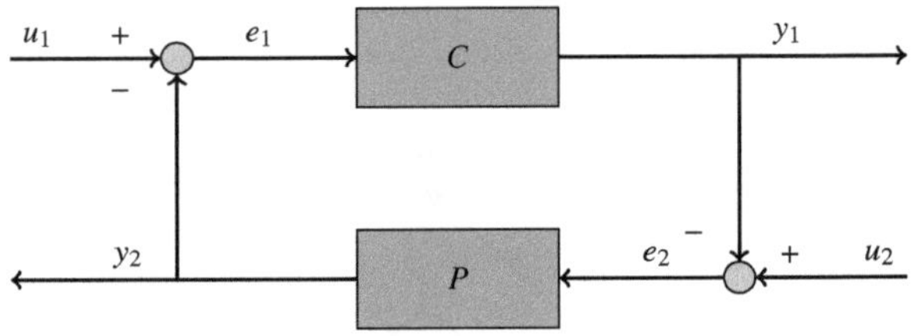

Fig. 6.3 Closed-loop system

tracker to the reference signal $\widehat{y}_1$. If we set $\widehat{u}_2 := \frac{s}{F}\,\widehat{\theta} \approx s\,\widehat{\theta}$, where $\widehat{u}_2$ is the target speed, then (6.1), and (6.2) yield:

$$\begin{cases} \widehat{\varepsilon} = \widehat{\theta} - \widehat{x} = \frac{F}{s}\,(\widehat{u}_2 - \widehat{y}_1), \\ \widehat{e}_1 = \widehat{u}_1 - G\,\widehat{\varepsilon}. \end{cases} \tag{6.3}$$

The main goal is to design a stabilizing controller C' such that $\lim_{t \to +\infty} \varepsilon(t) = 0$ for the above scenarios. To do that, let us first review results on stabilizability.

6.2 Parametrizations of all Stabilizing Controllers

Within the *fractional representation approach* to analysis and synthesis problems [1, 10], the class of systems that are considered are defined by transfer matrices with entries in the quotient field $Q(A) := \{n/d \mid 0 \neq d,\ n \in A\}$ of an integral domain A of SISO stable plants. Integral domains commonly considered are the Hardy algebra $H_\infty(\mathbb{C}_+)$ of bounded holomorphic functions in $\mathbb{C}_+ := \{s \in \mathbb{C} \mid \Re(s) > 0\}$, $RH_\infty := \mathbb{R}(s) \cap H_\infty(\mathbb{C}_+)$, the Wiener algebras $\widehat{\mathscr{A}}$ or W_+, the disc algebra $A(\mathbb{D})$, … For more details, see [1, 10]. Let us recall a few standard definitions [1, 10].

Definition 1 Let A be an integral domain of stable SISO plants and $K := Q(A)$.

- A *fractional representation* of the transfer matrix $P \in K^{q \times r}$ is any representation of the form $P = D^{-1} N = \widetilde{N} \widetilde{D}^{-1}$, where $D \in A^{q \times q}$, $\det D \neq 0$, $N \in A^{q \times r}$, $\widetilde{N} \in A^{q \times r}$, $\widetilde{D} \in A^{r \times r}$ and $\det \widetilde{D} \neq 0$.
- The plant $P \in K^{q \times r}$ is said to be *(internally) stabilizable* if there exists a *stabilizing controller* $C \in K^{r \times q}$ of P, namely a controller $C \in K^{r \times q}$ such that

$$H(P,\, C) := \begin{pmatrix} I_q & P \\ C & I_r \end{pmatrix}^{-1} \in A^{(q+r) \times (q+r)},$$

where, using Fig. 6.3, $(e_1^T \quad e_2^T)^T = H(P,\, C)\,(u_1^T \quad u_2^T)^T$ is defined by:

$$H(P,\ C) = \begin{pmatrix} (I_q - P\,C)^{-1} & -(I_q - P\,C)^{-1}\,P \\ -C\,(I_q - P\,C)^{-1} & I_r + C\,(I_q - P\,C)^{-1}\,P \end{pmatrix} = \begin{pmatrix} I_q + P\,(I_r - C\,P)^{-1}\,C & -P\,(I_r - C\,P)^{-1} \\ -(I_r - C\,P)^{-1}\,C & (I_r - C\,P)^{-1} \end{pmatrix}. \tag{6.4}$$

- A transfer matrix $P \in K^{q\times r}$ admits a *left-coprime factorization* if there exist $D \in A^{q\times q}$, $\det D \neq 0$, $N \in A^{q\times r}$, $X \in A^{q\times q}$ and $Y \in A^{r\times q}$ such that $P = D^{-1}\,N$ and $D\,X - N\,Y = I_q$.
- A transfer matrix $P \in K^{q\times r}$ admits a *right-coprime factorization* if there exist $\widetilde{D} \in A^{r\times r}$, $\det \widetilde{D} \neq 0$, $\widetilde{N} \in A^{q\times r}$, $\widetilde{Y} \in A^{r\times q}$ and $\widetilde{X} \in A^{r\times r}$ such that $P = \widetilde{N}\,\widetilde{D}^{-1}$ and $-\widetilde{Y}\,\widetilde{N} + \widetilde{X}\,\widetilde{D} = I_r$.
- A transfer matrix $P \in K^{q\times r}$ admits a *doubly coprime factorization* if P admits a left- and a right-coprime factorizations $P = D^{-1}\,N = \widetilde{N}\,\widetilde{D}^{-1}$ such that:

$$\begin{pmatrix} D & -N \\ -\widetilde{Y} & \widetilde{X} \end{pmatrix} \begin{pmatrix} X & \widetilde{N} \\ Y & \widetilde{D} \end{pmatrix} = I_{q+r}. \tag{6.5}$$

In what follows, we characterize stabilizablity and give a parametrization of all stabilizing controllers obtained in [8]. Contrary to [7, 8], which are based on modern algebraic methods, following [6], we present here elementary proofs of these results.

Proposition 1 ([7]) *Let $P \in K^{q\times r}$ be a plant, $M := (I_q \quad -P) \in K^{q\times(q+r)}$ and $\widetilde{M} := (P^T \quad I_r^T)^T \in K^{(q+r)\times r}$. Then, $P \in K^{q\times r}$ is stabilizable iff one of the following equivalent assertions is satisfied*:

1. *There exists a matrix $L := (S_o^T \quad U^T)^T \in A^{(q+r)\times q}$, $S_o \in A^{q\times q}$, $\det S_o \neq 0$, $U \in A^{r\times q}$, such that*:

   a. $L\,P = \begin{pmatrix} S_o\,P \\ U\,P \end{pmatrix} \in A^{(q+r)\times r}$,

   b. $M\,L = S_o - P\,U = I_q$.

   Then $C = U\,S_o^{-1}$ stabilizes P, $S_o = (I_q - P\,C)^{-1}$ is the output sensitivity transfer matrix and $U = C\,(I_q - P\,C)^{-1}$.
2. *There exists a matrix $\widetilde{L} := (-\widetilde{U} \quad \widetilde{S}_i) \in A^{r\times(q+r)}$, $\widetilde{U} \in A^{r\times q}$, $\widetilde{S}_i \in A^{r\times r}$, $\det \widetilde{S}_i \neq 0$, such that:*

   a. $P\,\widetilde{L} = (-P\,\widetilde{U} \quad P\,\widetilde{S}_i) \in A^{q\times(q+r)}$,

   b. $\widetilde{L}\,\widetilde{M} = -\widetilde{U}\,P + \widetilde{S}_i = I_r$.

   Then $\widetilde{C} = \widetilde{S}_i^{-1}\,\widetilde{U}$ stabilizes P, $\widetilde{S}_i = (I_r - \widetilde{C}\,P)^{-1}$ is the input sensitivity transfer matrix and $\widetilde{U} = (I_r - \widetilde{C}\,P)^{-1}\,\widetilde{C}$

With the above notations, we have $\widetilde{C} = C \Leftrightarrow \widetilde{L}\,L = 0$.

Proof 1. Let C stabilize P. Using (6.4), we get $S_o := (I_q - P\,C)^{-1} \in A^{q\times q}$ and $U := C\,(I_q - P\,C)^{-1} \in A^{r\times q}$. Using again (6.4), $L := (S_o^T \quad U^T)^T \in A^{(q+r)\times q}$ satisfies $L\,P \in A^{(q+r)\times r}$, and thus 1.a. holds. Finally, 1.b. also holds since we have

$S_o - P\,U = (I_q - P\,C)^{-1} - P\,C\,(I_q - P\,C)^{-1} = I_q$. Let us now suppose that 1.a and 1.b hold. Since $\det S_o \neq 0$, $S_o - P\,U = I_q$ yields $I_q - P\,(U\,S_o^{-1}) = S_o^{-1}$, i.e., with the notation $C := U\,S_o^{-1}$, we get $S_o = (I_q - P\,C)^{-1}$. Then, we have $(I_q - P\,C)^{-1} \in A^{q\times q}$ and $U = C\,S_o = C\,(I_q - P\,C)^{-1} \in A^{r\times q}$. Now, using $L\,P \in A^{(q+r)\times r}$, we get $(I_q - P\,C)^{-1}\,P \in A^{q\times r}$ and $C\,(I_q - P\,C)^{-1}\,P \in A^{r\times r}$, and thus $H(P, C) \in A^{(q+r)\times(q+r)}$, i.e., C stabilizes P. 2 can be proved similarly. Finally, we have $\widetilde{C} = C$ iff $-\widetilde{U}\,S_o + \widetilde{S}_i\,U = 0$, i.e., iff $\widetilde{L}\,L = 0$. □

Corollary 1 ([7]) *P is stabilizable iff there exists $U \in A^{r\times q}$ such that:*

$$\begin{cases} T_i := -U\,P \in A^{r\times r}, \\ T_o := -P\,U \in A^{q\times q}, \\ R := (I_q + P\,U)\,P = P\,(I_r + U\,P) \in A^{q\times r}, \\ \det(I_q + P\,U) = \det(I_r + U\,P) \neq 0. \end{cases} \tag{6.6}$$

Then, $C := U\,(P\,U + I_q)^{-1} = (U\,P + I_r)^{-1}\,U$ is a stabilizing controller of P,

$$\text{Feedback}(C, P) := U = C\,(I_q - P\,C)^{-1} = (I_r - C\,P)^{-1}\,C, \tag{6.7}$$

T_i (resp., T_o) is the complementary input (resp. output) sensitivity transfer matrix.

Proof By 1 of Proposition 1, P is stabilizable iff there exists $U \in A^{r\times q}$ such that $S_o = I_q + P\,U \in A^{q\times q}$, $\det S_o \neq 0$, $S_o\,P \in A^{q\times r}$ and $U\,P \in A^{r\times r}$, i.e., iff $-T_o := P\,U \in A^{q\times q}$, $R := (I_q + P\,U)\,P = P\,(I_r + U\,P) \in A^{q\times r}$, $-T_i := U\,P \in A^{r\times r}$ and $\det(I_q + P\,U) = \det(I_r + U\,P) \neq 0$.

If C stabilizes P, then the matrices L and $\widetilde{L}$ defined by

$$\begin{cases} S_o := (I_q - P\,C)^{-1}, \\ U = \widetilde{U} := C\,(I_q - P\,C)^{-1} = (I_r - C\,P)^{-1}\,C, \\ S_i = \widetilde{S}_i := (I_r - C\,P)^{-1}, \end{cases}$$

satisfy 1 and 2 of Proposition 1 and $R = S_o\,P = P\,S_i$, 1.b and 2.b show that

$$\Pi_C := L\,M = \begin{pmatrix} S_o & -R \\ U & T_i \end{pmatrix}, \quad \Pi_P := \widetilde{M}\,\widetilde{L} = \begin{pmatrix} T_o & R \\ -U & S_i \end{pmatrix}$$

are two *complementary projectors* of $A^{(q+r)\times(q+r)}$, i.e., $\Pi_C^2 = \Pi_C$, $\Pi_P^2 = \Pi_P$ and $\Pi_C + \Pi_P = I_{q+r}$. Using (6.4), Fig. 6.3 and $T_i = -C\,R$, we can easily check that:

$$\begin{pmatrix} e_1 \\ y_1 \end{pmatrix} = \Pi_C \begin{pmatrix} u_1 \\ u_2 \end{pmatrix}, \quad \begin{pmatrix} y_2 \\ e_2 \end{pmatrix} = \Pi_P \begin{pmatrix} u_1 \\ u_2 \end{pmatrix}.$$

Corollary 2 *Let $\Delta \in A^{q \times r}$. The following assertions are equivalent*:

1. $C \in K^{r\times q}$ *stabilizes* $P \in K^{q\times r}$.
2. $C' := C\,(I_q - \Delta\, C)^{-1} = (I_r - C\,\Delta)^{-1}\, C$ *stabilizes* $P' := P - \Delta$.

Proof By Corollary 1, C stabilizes P iff $U := C\,(I_q - P\,C)^{-1} = (I_r - C\,P)^{-1}\,C$ is such that $U \in A^{r\times q}$ and satisfies (6.6). Now, for any $\Delta \in A^{q\times r}$, $U \in A^{r\times q}$ and (6.6) are equivalent to $U \in A^{r\times q}$ and

$$\begin{cases} U\,(P-\Delta) = U\,P - U\,\Delta \in A^{r\times r}, \\ (P-\Delta)\,U = P\,U - \Delta\,U \in A^{q\times q}, \\ ((P-\Delta)\,U + I_q)\,(P-\Delta) = (P\,U+I_q)\,P - (P\,U+I_q)\,\Delta - \Delta\,(U\,P) + \Delta\,U\,\Delta \in A^{q\times r}, \end{cases}$$

which shows that $P \in K^{q\times r}$ is stabilized by $C \in K^{r\times q}$ iff $P - \Delta$ is stabilized by $C' := U\,((P-\Delta)\,U + I_q)^{-1} = (U\,(P-\Delta) + I_r)^{-1}\,U$. Finally, using $C = U\,(P\,U + I_q)^{-1} = (U\,P + I_r)^{-1}\,U$, we obtain:

$$\begin{aligned} C' &= U\,(P\,U + I_q - \Delta\,U)^{-1} = U\,((I_q - \Delta\,U\,(P\,U+I_q)^{-1})\,(P\,U+I_q))^{-1} \\ &= U\,(P\,U+I_q)^{-1}\,(I_q - \Delta\,U\,(P\,U+I_q)^{-1})^{-1} = C\,(I_q - \Delta\,C)^{-1}, \end{aligned}$$

$$\begin{aligned} C' &= (U\,P + I_r - U\,\Delta)^{-1}\,U = ((U\,P+I_r)\,(I_r - (U\,P+I_r)^{-1}\,U\,\Delta))^{-1}\,U \\ &= (I_r - (U\,P+I_r)^{-1}\,U\,\Delta))^{-1}\,(U\,P+I_r)^{-1}\,U = (I_r - C\,\Delta)^{-1}\,C. \end{aligned}$$ □

The next results gives a parametrization of all the stabilizing controllers of a stabilizable plant. Only the explicit knowledge of a stabilizing controller is assumed.

Theorem 1 ([8]) *Let $C_\star \in K^{r\times q}$ be a stabilizing controller of $P \in K^{q\times r}$ and*:

$$\begin{aligned} &U := C_\star\,(I_q - P\,C_\star)^{-1} = (I_r - C_\star\,P)^{-1}\,C_\star \in A^{r\times q}, \\ &S_i := (I_r - C_\star\,P)^{-1} \in A^{r\times r}, \quad S_o := (I_q - P\,C_\star)^{-1} \in A^{q\times q}. \end{aligned}$$

Then, all stabilizing controllers of P are given by

$$C(\Lambda) = (U + \Lambda)\,(S_o + P\,\Lambda)^{-1} = (S_i + \Lambda\,P)^{-1}\,(U + \Lambda), \tag{6.8}$$

where Λ is any matrix which belongs to the following A-module

$$\Omega = \{\,\Lambda \in A^{r\times q} \mid \Lambda\,P \in A^{r\times r},\ P\,\Lambda \in A^{q\times q},\ P\,\Lambda\,P \in A^{q\times r}\,\}, \tag{6.9}$$

and satisfies:

$$\det(S_o + P\,\Lambda) \neq 0, \quad \det(S_i + \Lambda\,P) \neq 0. \tag{6.10}$$

Proof Let C_1 and $C_2 \in K^{r\times q}$ be two stabilizing controllers of $P \in K^{q\times r}$ and:

$$\begin{aligned} &S_{ik} := (I_r - C_k\,P)^{-1} \in A^{r\times r}, \quad S_{ok} := (I_q - P\,C_k)^{-1} \in A^{q\times q}, \\ &U_k := C_k\,(I_q - P\,C_k)^{-1} = (I_r - C_k\,P)^{-1}\,C_k := \widetilde{U}_k \in A^{r\times q}, \end{aligned} \quad k = 1, 2.$$

We then have $C_k = U_k\, S_{ok}^{-1} = S_{ik}^{-1}\, U_k$ for $k = 1, 2$. Moreover, the matrices $L_k := (S_{ok}^T \quad U_k^T)^T \in A^{(q+r)\times q}$ and $\widetilde{L}_k := (-U_k \quad S_{ik}) \in A^{r\times(q+r)}$ satisfy 1.a, 1.b, 2.a and 2.b of Proposition 1. Using 1.b and 2.b, we get:

$$\begin{cases} S_{o2} - S_{o1} = P\,U_2 + I_q - P\,U_1 - I_q = P\,(U_2 - U_1), \\ S_{i2} - S_{i1} \;= U_2\,P + I_r - U_1\,P - I_r = (U_2 - U_1)\,P. \end{cases}$$

Now, using 1.a and 2.a, we obtain:

$$\begin{cases} (U_2 - U_1)\,P \quad\;\; = S_{i2} - S_{i1} \in A^{r\times r}, \\ P\,(U_2 - U_1) \quad\;\; = S_{o2} - S_{o1} \in A^{q\times q}, \\ P\,(U_2 - U_1)\,P = (S_{o2} - S_{o1})\,P = P\,(S_{i2} - S_{i1}) \in A^{q\times r}, \end{cases} \Rightarrow\; U_2 - U_1 \in \Omega.$$

Hence, if $\Lambda := U_2 - U_1 \in \Omega$, then we have $U_2 = U_1 + \Lambda$, $S_{o2} = S_{o1} + P\,\Lambda$ and $S_{i2} = S_{i1} + \Lambda\,P$, and if $\det(S_{o1} + P\,\Lambda) \neq 0$ and $\det(S_{i1} + \Lambda\,P) \neq 0$, then we get:

$$C_2 = U_2\,S_{o2}^{-1} = (U_1 + \Lambda)\,(S_{o1} + P\,\Lambda)^{-1}, \; C_2 = S_{i2}^{-1}\,U_2 = (S_{i1} + \Lambda\,P)^{-1}(U_1 + \Lambda).$$

If $S_o := S_{o1}$, $U := U_1$ and $S_i := S_{i1}$, then we get $C_2 = C(\Lambda)$, where $C(\Lambda)$ is defined by (6.8) for a certain $\Lambda \in \Omega$ which satisfies (6.10). Finally, let us prove that for every $\Lambda \in \Omega$ which satisfies (6.10), the controller $C(\Lambda)$ defined by (6.8) stabilizes P. Let $L(\Lambda) := ((S_o + P\,\Lambda)^T \quad (U + \Lambda)^T)^T$, $\widetilde{L}(\Lambda) := (-(U + \Lambda) \quad S_i + \Lambda\,P)$, $M := (I_q \quad -P)$ and $\widetilde{M} := (P^T \quad I_r^T)^T$. Since $\Lambda \in \Omega$, we have $U + \Lambda \in A^{r\times q}$, $S_o + P\,\Lambda \in A^{q\times q}$ and

$$L(\Lambda)\,P = \begin{pmatrix} S_o\,P + P\,\Lambda\,P \\ U\,P + \Lambda\,P \end{pmatrix} \in A^{(q+r)\times r}, \quad M\,L(\Lambda) = S_o - P\,U = I_q,$$

and thus $C(\Lambda) = (U + \Lambda)\,(S_o + P\,\Lambda)^{-1}$ stabilizes P by 1 of Proposition 1. Similarly, since $\Lambda \in \Omega$, we get $U + \Lambda \in A^{r\times q}$, $S_i + \Lambda\,P \in A^{r\times r}$ and

$$\begin{cases} P\,\widetilde{L}(\Lambda) = (-(P\,U + P\,\Lambda) \quad P\,S_i + P\,\Lambda\,P) \in A^{q\times(q+r)}, \\ \widetilde{L}(\Lambda)\,\widetilde{M} = -U\,P + S_i = I_r, \end{cases}$$

i.e., $C(\Lambda) = (S_i + \Lambda\,P)^{-1}\,(U + \Lambda)$ stabilizes P by 2 of of Proposition 1. □

Proposition 2 ([8]) *Let $C_\star \in K^{r\times q}$ be a stabilizing controller of $P \in K^{q\times r}$ and:*

$$\begin{cases} L = \begin{pmatrix} (I_q - P\,C_\star)^{-1} \\ C_\star\,(I_q - P\,C_\star)^{-1} \end{pmatrix} \in A^{(q+r)\times q}, \\ \widetilde{L} = (-(I_r - C_\star\,P)^{-1}\,C_\star \quad (I_r - C_\star\,P)^{-1}) \in A^{r\times(q+r)}. \end{cases}$$

Then, the A-module Ω defined by (6.9) *satisfies* $\Omega = \widetilde{L}\,A^{(q+r)\times(q+r)}\,L$. *Hence, the A-module Ω is generated by the $(q+r)^2$ matrices $\widetilde{L}_{\bullet i}\,L_{j\bullet}$, where $\widetilde{L}_{\bullet i}$ is the ith column of $\widetilde{L}$ and $L_{j\bullet}$ the jth row of L, i.e.,* $\Omega = \left\{\sum_{i,j=1}^{q+r} a_{ij}\,\widetilde{L}_{\bullet i}\,L_{j\bullet} \mid a_{ij} \in A\right\}$.

Proof Let $\Lambda \in \Omega$, i.e., $\Lambda \in A^{r\times q}$ is such that $\Lambda P \in A^{r\times r}$, $P\Lambda \in A^{q\times q}$ and $P\Lambda P \in A^{q\times r}$, $M := (I_q - P)$ and $\widetilde{M} := (P^T \quad I_r^T)^T$. In particular, we have $\widetilde{M}\,\Lambda \in A^{(q+r)\times q}$, $\Lambda M \in A^{r\times(q+r)}$ and $P\,\Lambda\,M \in A^{q\times(q+r)}$. Now, using 2.b of Proposition 1, i.e., $\widetilde{L}\,\widetilde{M} = I_r$, we get $\Lambda = \widetilde{L}\,(\widetilde{M}\,\Lambda)$. Using 1.b of Proposition 1, i.e., $M\,L = I_q$, we get $\Lambda = (\Lambda\,M)\,L$, and thus $P\,\Lambda = (P\,\Lambda\,M)\,L$. Substituting $\Lambda = (\Lambda\,M)\,L$ and $P\,\Lambda = (P\,\Lambda\,M)\,L$ into $\widetilde{M}\,\Lambda = ((P\,\Lambda)^T \quad \Lambda^T)^T$, we get $\widetilde{M}\,\Lambda = \Theta\,L$, where $\Theta := ((P\,\Lambda\,M)^T (\Lambda\,M)^T)^T \in A^{(q+r)\times(q+r)}$ and, since $\Lambda = \widetilde{L}\,(\widetilde{M}\,\Lambda)$, we finally obtain $\Lambda = \widetilde{L}\,\Theta\,L$, i.e., $\Omega \subseteq \widetilde{L}\,A^{(q+r)\times(q+r)}\,L$.

Now, let $\Lambda \in \widetilde{L}\,A^{(q+r)\times(q+r)}\,L$, i.e., $\Lambda = \widetilde{L}\,\Theta\,L$ for $\Theta \in A^{(q+r)\times(q+r)}$ and L and $\widetilde{L}$ satisfy 1 and 2 of Proposition 1. Then, using 1.a and 2.a of Proposition 1, we obtain $\Lambda \in A^{r\times q}$, $\Lambda\,P = \widetilde{L}\,\Theta\,(L\,P) \in A^{r\times r}$, $P\,\Lambda = (P\,\widetilde{L})\,\Theta\,L \in A^{q\times q}$, $P\,\Lambda\,P = (P\,\widetilde{L})\,\Theta\,(L\,P) \in A^{q\times r}$, i.e., $\Lambda \in \Omega$, and thus $\Omega = \widetilde{L}\,A^{(q+r)\times(q+r)}\,L$.

Finally, $\Theta \in A^{(q+r)\times(q+r)}$ can be written as $\Theta = \sum_{i,j=1}^{q+r} \Theta_{ij}\,E_{ij}$ where $\Theta_{ij} \in A$ and E_{ij} is the matrix defined by 1 in the ith row and the jth column and 0 elsewhere, and thus every $\Lambda \in \Omega$ can be written as $\Lambda = \widetilde{L}\,\Theta\,L = \sum_{i,j=1}^{q+r} \Theta_{ij}\,(\widetilde{L}\,E_{ij}\,L)$. Therefore, $\{\widetilde{L}\,E_{ij}\,L\}_{i,j=1,\ldots,q+r}$ is a family of generators of the A-module Ω and $\widetilde{L}\,E_{ij}\,L$ is the product of the ith column $\widetilde{L}_{\bullet i}$ of $\widetilde{L}$ by the jth row $L_{j\bullet}$ of L. □

Combining Theorem 1 and Proposition 2, we obtain the following result.

Corollary 3 ([8]) *With the notations of Theorem 1, if $C_\star \in K^{r\times q}$ is a stabilizing controller of $P \in K^{q\times r}$, then all the stabilizing controllers of P are of the form* (6.8), *where Λ is any matrix which belongs to $\Omega = \sum_{i,j=1}^{q+r} A\,(\widetilde{L}_{\bullet i}\,L_{j\bullet})$ and satisfies* (6.10).

In Corollary 3, only the explicit knowledge of a stabilizing controller is assumed. For many classes of infinite-dimensional systems, (PID, finite-dimensional) stabilizing controllers are known which is not the case for doubly coprime factorizations.

Corollary 4
1. *If $P \in K^{q\times r}$ admits a left-coprime factorization $P = D^{-1}\,N$, $D\,X - N\,Y = I_q$, with $(X^T \quad Y^T)^T \in A^{(q+r)\times q}$ and* $\det X \neq 0$, *then the matrix $L = ((X\,D)^T \quad (Y\,D)^T)^T \in A^{(q+r)\times q}$ satisfies 1.a and 1.b of Proposition 1, and $C = Y\,X^{-1}$ is a stabilizing controller of P.*
2. *If $P \in K^{q\times r}$ admits a right-coprime factorization $P = \widetilde{N}\,\widetilde{D}^{-1}$, $-\widetilde{Y}\,\widetilde{N} + \widetilde{X}\,\widetilde{D} = I_r$, with $(-\widetilde{Y} \quad \widetilde{X}) \in A^{r\times(q+r)}$ and* $\det \widetilde{X} \neq 0$, *then the matrix $\widetilde{L} = (-\widetilde{D}\,\widetilde{Y} \quad \widetilde{D}\,\widetilde{X}) \in A^{r\times(q+r)}$ satisfies 2.a and 2.b of Proposition 1, and $C = \widetilde{X}^{-1}\,\widetilde{Y}$ is a stabilizing controller of P.*

Proof Let us prove 1. If $P = D^{-1}\,N$, $D\,X - N\,Y = I_q$, is a left-coprime factorization of P, then $(X\,D)\,P = X\,N \in A^{q\times r}$, $(Y\,D)\,P = Y\,N \in A^{r\times r}$ and $D\,X - N\,Y = I_q \Rightarrow X - P\,Y = D^{-1} \Rightarrow (X\,D) - P\,(Y\,D) = I_q$, i.e., $L = ((X\,D)^T \quad (Y\,D)^T)^T \in A^{(q+r)\times q}$ satisfies 1 of Proposition 1, and thus $C = (Y\,D)\,(X\,D)^{-1} = Y\,X^{-1}$ stabilizes P. 2 can be proved similarly. □

From Corollary 4, the existence of a doubly coprime factorization of P is a sufficient but not a necessary condition for stabilizability.

Proposition 3 ([8]) *If* $P \in K^{q\times r}$ *admits the doubly coprime factorization* (6.5), *then the A-module* Ω *defined by* (6.9) *satisfies* $\Omega = \widetilde{D}\, A^{r\times q}\, D$.

Proof Let $\Lambda \in \widetilde{D}\, A^{r\times q}\, D$, i.e., $\Lambda = \widetilde{D}\, Q\, D$ for a certain $Q \in A^{r\times q}$. Then, we have $\Lambda = \widetilde{D}\, Q\, D \in A^{r\times q}$, $\Lambda\, P = \widetilde{D}\, Q\, N \in A^{r\times r}$, $P\, \Lambda = \widetilde{N}\, Q\, D \in A^{q\times q}$ and $P\, \Lambda\, P = \widetilde{N}\, Q\, N \in A^{q\times r}$, which shows that $\Lambda \in \Omega$. Conversely, let $\Lambda \in \Omega$ and $Q := \widetilde{D}^{-1}\, \Lambda\, D^{-1} \in K^{r\times q}$. From (6.5), we get the identities $D^{-1} = X - P\, Y$ and $\widetilde{D}^{-1} = \widetilde{X} - \widetilde{Y}\, P$, which yield $Q = \widetilde{D}^{-1}\, \Lambda\, D^{-1} = (\widetilde{X} - \widetilde{Y}\, P)\, \Lambda\, (X - P\, Y) = \widetilde{X}\, \Lambda\, X - \widetilde{X}\, (\Lambda\, P)\, Y - \widetilde{Y}\, (P\, \Lambda)\, X + \widetilde{Y}\, (P\, \Lambda\, P)\, Y \in A^{r\times q}$ since $\Lambda \in \Omega$ and the entries of X, Y, $\widetilde{X}$ and $\widetilde{Y}$ belong to A. Therefore, we get $\Lambda = \widetilde{D}\, Q\, D$ for a certain $Q \in A^{r\times q}$, i.e., $\Lambda \in \widetilde{D}\, A^{r\times q}\, D$, which finally proves that $\Omega = \widetilde{D}\, A^{r\times q}\, D$. □

The next corollary shows that the parametrization (6.8) gives rise to the Youla-Kučera parametrization when the plant P admits a doubly coprime factorization.

Corollary 5 ([8]) *Let* $P \in K^{q\times r}$ *admit a doubly coprime factorization* $P = D^{-1}\, N = \widetilde{N}\, \widetilde{D}^{-1}$, *where* (6.5) *is satisfied. Then, all the stabilizing controllers of P are of the form* $C(Q) = (Y + \widetilde{D}\, Q)\, (X + \widetilde{N}\, Q)^{-1} = (\widetilde{X} + Q\, N)^{-1}\, (\widetilde{Y} + Q\, D)$, *where* Q *is any matrix of* $A^{r\times q}$ *such that* $\det(X + \widetilde{N}\, Q) \neq 0$ *and* $\det(\widetilde{X} + Q\, N) \neq 0$.

Proof By Proposition 3, we have $\Omega = \widetilde{D}\, A^{r\times q}\, D$. Moreover, by 1 of Corollary 4, $C = (Y\, D)\, (X\, D)^{-1} = Y\, X^{-1}$ is a stabilizing controller of P. Moreover, by 2 of Corollary 4, $\widetilde{C} = (\widetilde{D}\, \widetilde{X})^{-1}\, (\widetilde{D}\, \widetilde{Y}) = \widetilde{X}^{-1}\, \widetilde{Y}$ is a stabilizing controller of P. By (6.5), $-\widetilde{Y}\, X + \widetilde{X}\, Y = 0$, which shows that $\widetilde{C} = C$. Therefore, by Theorem 1 or Corollary 3, we obtain that all the stabilizing controllers of P are of the form

$$\begin{aligned} C^{\circ}(Q) &:= C(\widetilde{D}\, Q\, D) = (Y\, D + \widetilde{D}\, Q\, D)\, (X\, D + P\, \widetilde{D}\, Q\, D)^{-1} \\ &= (Y\, D + \widetilde{D}\, Q\, D)\, (X\, D + \widetilde{N}\, Q\, D)^{-1} = (Y + \widetilde{D}\, Q)\, D\, D^{-1}\, (X + \widetilde{N}\, Q)^{-1} \\ &= (Y + \widetilde{D}\, Q)\, (X + \widetilde{N}\, Q)^{-1}, \end{aligned}$$

$$\begin{aligned} C^{\circ}(Q) &:= C(\widetilde{D}\, Q\, D) = (\widetilde{D}\, \widetilde{X} + \widetilde{D}\, Q\, D\, P)^{-1}\, (\widetilde{D}\, \widetilde{Y} + \widetilde{D}\, Q\, D) \\ &= (\widetilde{D}\, \widetilde{X} + \widetilde{D}\, Q\, N)^{-1}\, (\widetilde{D}\, \widetilde{Y} + \widetilde{D}\, Q\, D) = (\widetilde{X} + Q\, N)^{-1}\, \widetilde{D}^{-1}\, \widetilde{D}\, (\widetilde{Y} + Q\, D) \\ &= (\widetilde{X} + Q\, N)^{-1}\, (\widetilde{Y} + Q\, D), \end{aligned}$$

where $Q \in A^{r\times q}$ is any matrix such that $\det(X + \widetilde{N}\, Q) \neq 0$ and $\det(\widetilde{X} + Q\, N) \neq 0$. □

The following result is a direct consequence of Corollaries 3 and 2.

Theorem 2 *Let* $\Delta \in A^{q\times r}$ *and* $C_\star \in K^{r\times q}$ *be a stabilizing controller of the plant* $P \in K^{q\times r}$. *Then, all the stabilizing controllers* C' *of* $P' := P - \Delta$ *are of the form*

$$C'(\Lambda) = C(\Lambda)\, (I_q - \Delta\, C(\Lambda))^{-1} = (I_r - C(\Lambda)\Delta)^{-1} C(\Lambda) = \text{Feedback}\, (C(\Lambda), \Delta)\,,$$

where $C(\Lambda)$ *is the parametrization* (6.8) *of all the stabilizing controllers of P.*

We have the following straightforward consequence of Theorem 2.

Corollary 6 *Let $P \in K^{q\times r}$ admits a doubly coprime factorization $P = D^{-1} N = \widetilde{N}\,\widetilde{D}^{-1}$, where (6.5) is satisfied, and $\Delta \in A^{q\times r}$. Then, $P' := P - \Delta$ admits the doubly coprime factorization $P' = D^{-1}\,(N - D\,\Delta) = (\widetilde{N} - \Delta\,\widetilde{D})\,\widetilde{D}^{-1}$ and:*

$$\begin{pmatrix} D & -(N - D\,\Delta) \\ -\widetilde{Y} & \widetilde{X} - \widetilde{Y}\,\Delta \end{pmatrix} \begin{pmatrix} X - \Delta\,Y & \widetilde{N} - \Delta\,\widetilde{D} \\ Y & \widetilde{D} \end{pmatrix} = I_{q+r}.$$

Hence, if $C(Q) := (Y + \widetilde{D}\,Q)\,(X + \widetilde{N}\,Q)^{-1} = (\widetilde{X} + Q\,N)^{-1}\,(\widetilde{Y} + Q\,D)$ is the Youla-Kučera parametrization of all the stabilizing controllers of P, then Youla-Kučera parametrization $C'(Q)$ of all the stabilizing controllers of P' satisfies:

$$C'(Q) = C(Q)\,(I_q - \Delta\,C(Q))^{-1} = (I_r - C(Q)\,\Delta)^{-1}\,C(Q) = \text{Feedback}\,(C(Q), \Delta)\,. \tag{6.11}$$

Theorem 2 and Corollary 6 are particularly interesting when P is a rational transfer matrix for which different techniques can be used to find a particular finite-dimensional controller or doubly coprime factorization.

Example 1 Let $F \in A := H_\infty(\mathbb{C}_+)$ be such that $F_0 := F(0) \neq 0$, $P := \frac{F}{s}$, $\Delta := F\,\frac{(1-e^{-T\,s})}{s} \in A$, and $P' := P - \Delta = F\,\frac{e^{-T\,s}}{s}$. Clearly, P admits the coprime factorization $P = \frac{N}{D}$, $D\,X - N\,Y = 1$, where $\alpha \in \mathbb{R}_{>0} := \{x \in \mathbb{R} \mid x > 0\}$,

$$N = \frac{F}{s+\alpha}, \quad D = \frac{s}{s+\alpha}, \quad X = 1 + \alpha\,\frac{1 - \frac{F}{F_0}}{s}, \quad Y = -\frac{\alpha}{F_0}.$$

The only point to check is that $X \in A$, i.e., $Z := (X-1)/\alpha = \frac{1-\frac{F}{F_0}}{s} \in A$. Clearly, Z is a holomorphic function in $\mathbb{C}_+$, has no poles in the imaginary axis and

$$\left| \frac{1 - \frac{F(i\,\omega)}{F_0}}{i\,\omega} \right| \leq \frac{1 + \left|\frac{F(i\,\omega)}{F_0}\right|}{|\omega|} \leq \frac{1 + \parallel \frac{F}{F_0} \parallel_\infty}{|\omega|},$$

which proves that $Z \in A$. By Corollary 5, the Youla-Kučera parametrization of all the stabilizing controllers of P is then defined by:

$$\forall\, Q \in A:\; C(Q) = \frac{Y + D\,Q}{X + N\,Q} = \frac{-\frac{\alpha}{F_0} + \frac{s}{s+\alpha}\,Q}{1 + \alpha\,\frac{1-\frac{F}{F_0}}{s} + \frac{F}{s+\alpha}\,Q}. \tag{6.12}$$

By Corollary 6, the Youla-Kučera parametrization of $P' = F\,\frac{e^{-T\,s}}{s}$ is then

$$C'(Q) := \frac{C(Q)}{1 - \Delta\, C(Q)} = \frac{Y + D\,Q}{(X - \Delta\, Y) + (N - \Delta\, D)\,Q} = \frac{-\frac{\alpha}{F_0} + \frac{s}{s+\alpha}\,Q}{1 + \alpha\,\frac{1 - \frac{F}{F_0}\,e^{-T\,s}}{s} + Q\,F\,\frac{e^{-Ts}}{s+\alpha}}, \tag{6.13}$$

for the coprime factorization $P' = \frac{N - D\,\Delta}{D}$, $D\,(X - \Delta\,Y) - (N - D\,\Delta)\,Y = 1$.

Example 2 We can apply Corollary 6 to $P = \frac{1}{2\,e}\,\frac{s+1}{s-1}$ and $\Delta = \frac{1}{2\,e}\,\frac{(s+1) - 2\,e^{1-\sqrt{s}}}{s-1} \in A := H_\infty(\mathbb{C}_+)$ to get the Youla-Kučera parametrization of $P' := P - \Delta = \frac{e^{-\sqrt{s}}}{s-1}$.

6.3 Study of the Tracking Problem and Numerical Simulations

In this section, we study the tracking problems introduced in Sect. 6.1.

Let $F \in A := H_\infty(\mathbb{C}_+)$ be such that $F_0 := F(0) \neq 0$. In many situations, we have $F \in RH_\infty$. Using Fig. 6.4 and (6.3), let us introduce the following two systems:

$$\begin{cases} \widehat{\varepsilon} = P\,\widehat{e}_2, \\ \widehat{y}_2 = G\,\widehat{\varepsilon}, \end{cases} \quad \begin{cases} P := \frac{F}{s}, \\ G := e^{-Ts}, \end{cases} \quad \begin{cases} \widehat{e}_1 = \widehat{u}_1 - \widehat{y}_2, \\ \widehat{e}_2 = \widehat{u}_2 - \widehat{y}_1. \end{cases}$$

We then have $\widehat{y}_2 = P'\,\widehat{e}_2$, where $P' := P\,G = \frac{F\,e^{-Ts}}{s}$ was introduced in Example 1. Considering the controller $\widehat{y}_1 = C'\,\widehat{e}_1$, we then obtain:

$$\begin{pmatrix} \widehat{\varepsilon} \\ \widehat{y}_1 \end{pmatrix} = \frac{1}{1 - P'\,C'} \begin{pmatrix} -P\,C' & P \\ C' & -P'\,C' \end{pmatrix} \begin{pmatrix} \widehat{u}_1 \\ \widehat{u}_2 \end{pmatrix}. \tag{6.14}$$

Lemma 1 *With the above notations, the following assertions are equivalent*:

1. $C'_\star \in Q(A)$ *is such that* $\frac{P\,C'_\star}{1-P'\,C'_\star}, \frac{P}{1-P'\,C'_\star}, \frac{C'_\star}{1-P'\,C'_\star}, \frac{P'\,C'_\star}{1-P'\,C'_\star} \in A$.
2. $C'_\star \in Q(A)$ *stabilizes* P', *i.e.*, $\frac{1}{1-P'\,C'_\star}, \frac{C'_\star}{1-P'\,C'_\star}, \frac{P'}{1-P'\,C'_\star} \in A$.

If $C(Q)$ *is the Youla-Kučera parametrization* (6.12) *of* $P = \frac{F}{s}$, *then the Youla-Kučera parametrization of* $P' = P - \Delta$, *where* $\Delta := F\,\frac{(1-e^{-T\,s})}{s} \in A$, *is given by* (6.13).

Proof Let $C'_\star$ satisfy 1. Then, we have $\frac{1}{1-P'\,C'_\star} = \frac{P\,C'_\star}{1-P'\,C'_\star} - 1, \frac{C'_\star}{1-P'\,C'_\star} \in A$. We then get $\frac{P'}{1-P'\,C'_\star} = G\,\frac{P}{1-P'\,C'_\star} \in A$ since $G = e^{-Ts} \in A$, which proves 2. Now, let us suppose that $C'_\star$ stabilizes P', i.e., satisfies 2. We then need to check that $\frac{P}{1-P'\,C'_\star}, \frac{P\,C'_\star}{1-P'\,C'_\star} \in A$. Using (6.13), we have $C'_\star = C'(Q)$ for a certain $Q \in A$ and:

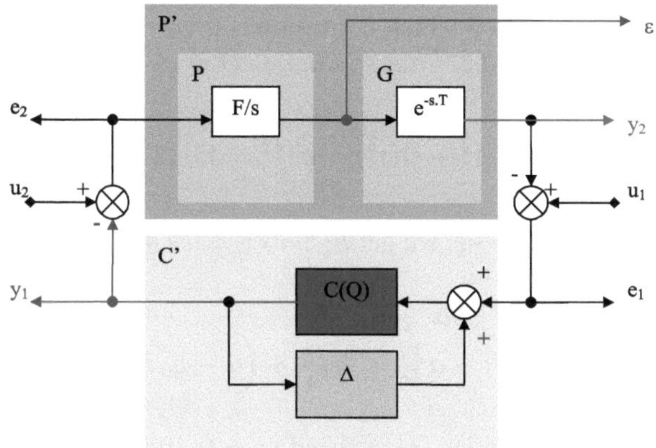

Fig. 6.4 Closed-loop system with the feedback structure of $C'(Q)$

$$\begin{cases} \frac{P}{1-P'\,C'_\star} = \frac{F}{s+\alpha}\left(1+\alpha\,\frac{1-\frac{F}{F_0}e^{-Ts}}{s} + Q\,F\,\frac{e^{-Ts}}{s+\alpha}\right) \in A, \\ \frac{P\,C'_\star}{1-P'\,C'_\star} = \frac{F}{s+\alpha}\left(-\frac{\alpha}{F_0} + \frac{s}{s+\alpha}\,Q\right) \in A. \end{cases}$$ □

The sensibility transfer function $S(Q) := (1 - P'\,C'(Q))^{-1}$ corresponding to the Youla-Kučera parametrization $C'(Q)$ of $P' = P - \Delta = F\,\frac{e^{-T\,s}}{s}$ (see (6.13)) is:

$$S(Q) = \frac{s}{s+\alpha}\left(1+\alpha\,\frac{1-\frac{F}{F_0}e^{-Ts}}{s} + Q\,F\,\frac{e^{-Ts}}{s+\alpha}\right).$$

Let us now investigate the asymptotic tracking of the target. In what follows, we consider the noiseless case, i.e., $\widehat{u}_1 = 0$. Using (6.14), we then get $\widehat{\varepsilon} = \frac{P}{1-P'\,C'}\,\widehat{u}_2$. Therefore, we have to find a stabilizing controller $C'(Q)$ of P' which is such that

$$\lim_{t\to+\infty} \varepsilon(t) = \lim_{s\to 0} s\,\widehat{\varepsilon}(s) = \lim_{s\to 0} s\,S(Q)\,P\,\widehat{u}_2 = 0,$$

where $\widehat{u}_2 = \frac{s}{F}\,\widehat{\theta}$ (see Sect. 6.1). Letting $\gamma = \theta_0^{(n)}$, $m = 1$ for **scenario 2** and $m = 2$ for **scenario 3**, $\widehat{u}_2$ can be decomposed as a sum of terms of the form $\widehat{u}_2 = \frac{\gamma}{s^n\,F}$, where $\gamma \in \mathbb{R}$ and $0 \leq n \leq m$. Using $s\,P = F$, we get $\lim_{t\to+\infty} \varepsilon(t) = \lim_{s\to 0} s\,S(Q)\,P\,\frac{\gamma}{s^n\,F} = \lim_{s\to 0} \frac{S(Q)\,\gamma}{s^n}$. Hence, if we set

$$E := \frac{(s+\alpha)^2\,S(Q)}{s^n} = \frac{(s+\alpha)\left(1+\alpha\,\frac{1-F/F_0\,e^{-T\,s}}{s}\right) + Q\,F\,e^{-T\,s}}{s^{n-1}},$$

then we have to determine the parameter $Q \in A$ such that $\lim_{s\to 0} E\,\frac{\gamma}{(s+\alpha)^2} = 0$.

Let us consider **scenario 3** which corresponds to $m = 2$. Let us consider power series expansion of Q, F and $e^{-T\,s}$ at $s = 0$, i.e., $Q = q_0 + q_1\,s + O(s^2)$,

$$F = F_0 + F_1\,s + \frac{F_2}{2}\,s^2 + O(s^3), \quad e^{-Ts} = 1 - T\,s + \frac{T^2}{2}\,s^2 + O(s^3).$$

where $F_i := F^{(i)}(0)$. For $n = 2$, we get $E_2 := s\,E = e_0 + e_1\,s + O(s^2)$, where:

$$\begin{cases} e_0 = q_0\,F_0 + \alpha + \alpha^2\,T - \alpha^2 \frac{F_1}{F_0}, \\ e_1 = 1 + q_1\,F_0 + q_0\,F_1 - \alpha \frac{F_1}{F_0} - \frac{\alpha^2\,F_2}{2\,F_0} + \left(\alpha - q_0\,F_0 + \alpha^2 \frac{F_1}{F_0}\right) T - \frac{\alpha^2\,T^2}{2}. \end{cases}$$

Hence, we have $E_2 = O(s^2)$, i.e., $e_0 = 0$ and $e_1 = 0$, iff:

$$\begin{cases} q_0 = -\dfrac{\alpha\,(1 + \alpha\,T)}{F_0} + \dfrac{\alpha^2}{F_0^2}\,F_1, \\ q_1 = -\dfrac{1 + 2\,\alpha\,T + \frac{\alpha^2\,T^2}{2}}{F_0} + \dfrac{\alpha(2 + \alpha\,T)\,F_1 + \alpha^2\,\frac{F_2}{2}}{F_0^2} - \dfrac{\alpha^2}{F_0^3}\,F_1^2. \end{cases}$$

For the numerical simulations, we take $Q := \frac{N_Q}{D_Q}$, where $N_Q := q_0 + q_2'\,s$, $D_Q := 1 + q_1'\,s$, q_0, $q_2' \in \mathbb{R}$, $q_1' \in \mathbb{R}_{\geq 0} = \{x \in \mathbb{R} \mid x \geq 0\}$, so that we get $Q = q_0 + (q_2' - q_0\,q_1')\,s + O(s^2)$, and thus we can choose arbitrarily $q_1' \in \mathbb{R}_{\geq 0}$ and:

$$\begin{cases} q_0 = -\dfrac{\alpha\,(1 + \alpha\,T)}{F_0} + \dfrac{\alpha^2}{F_0^2}\,F_1, \\ q_2' = -\dfrac{1 + \alpha\,q_1' + (\alpha\,q_1' + 2)\,\alpha\,T + \frac{\alpha^2\,T^2}{2}}{F_0} + \dfrac{((\alpha\,q_1' + 2)\,\alpha + \alpha^2\,T)\,F_1 + \alpha^2\,\frac{F_2}{2}}{F_0^2} - \dfrac{\alpha^2}{F_0^3}\,F_1^2. \end{cases}$$

We then have two degrees of freedom: $\alpha \in \mathbb{R}_{>0}$ and $q_1' \in \mathbb{R}_{\geq 0}$. We can check that we then have $\lim_{s \to 0} s\,S(Q)\,P\,\frac{\gamma}{s^n\,F} = 0$ for $0 \leq n \leq 1$. The form of Q can be used to study **scenario 2**, i.e., $Q = q_0 \in \mathbb{R}$, by considering $q_1' = 0$ and $q_2' = 0$.

For the visual tracking developed in Sect. 6.1, we have:

$$F := \frac{1 + \tau_2\,s}{(1 + \tau_1\,s)\,(1 + \tau_2\,s + \tau_2^2\,s^2)}, \quad \tau_1 = \frac{1}{60\,\pi}, \quad \tau_2 = \frac{1}{30\,\pi}, \quad T = 0.18.$$

In the Matlab simulations, we take $\alpha = 0.95$ and $q_1' = \frac{1}{5\pi}$ to get a gain margin of 5.4dB and a phase margin of 42° at 0.84Hz. See Black's diagram of the closed-loop (the blue plot in Fig. 6.5), the step response (the blue plot in Fig. 6.6) and compare with the results obtained with a PID controller (the black plots) and a H_∞-controller (the red plots).

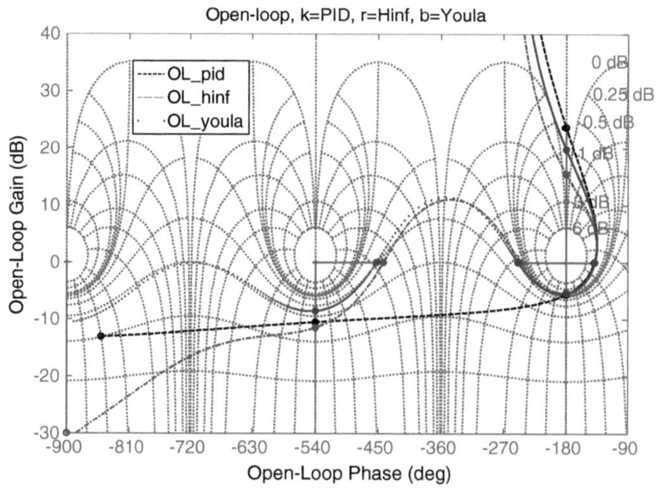

Fig. 6.5 Black's diagram and margins

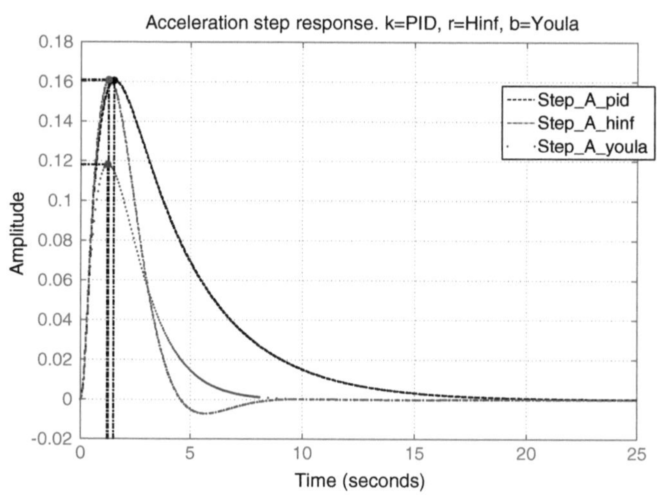

Fig. 6.6 Step response

References

1. Curtain, R.F., Zwart, H.J.: An Introduction to Infinite-Dimensional Linear Systems Theory. Texts in Mathematics, vol. 21, Springer, New York (1991)
2. Hilkert, J.M.: Inertially stabilized platforms technology. IEEE Control Syst. Mag. **28**(1), 26–46 (2008)
3. Hurák, Z., Řezáč, M.: Image-based pointing and tracking for inertially stabilized airborne camera platform. IEEE Trans. Control Syst. Technol. **99**, 1–14 (2011)

4. Masten, M.K.: Inertially stabilized platforms for optical imaging systems. IEEE Control Syst. Mag. **28**(1), 47–64 (2008)
5. Osborne, J., Hicks, G., Fuentes, R.: Global analysis of the double-gimbal mechanism. IEEE Control Syst. Mag. **28**, 44–64 (2008)
6. Quadrat, A.: An elementary proof of the general Q-parametrization of all stabilizing controllers. In: Proceedings of the 16th IFAC World Congress. Prague, Czech Republic (2005)
7. Quadrat, A.: A lattice approach to analysis and synthesis problems. Math. Control Sig. Syst. **18**, 147–186 (2006)
8. Quadrat, A.: On a generalization of the Youla-Kučera parametrization. Part II: the lattice approach to MIMO systems. Math. Control Sig. Syst. **18**, 199–235 (2006)
9. Quadrat, A., Quadrat, A.: Etude de l'effet du retard dans une boucle de poursuite d'un viseur gyrostabilisé. In: Proceedings of CIFA2012. Grenoble, 4–6 July 2012
10. Vidyasagar, M.: Control System Synthesis: A Factorization Approach. MIT Press, Cambridge (1985)

Part II
Numerical Methods

Chapter 7
Tuning an H-Infinity Controller with a Given Order and a Structure for Interconnected Systems with Delays

Suat Gumussoy and Wim Michiels

Abstract An eigenvalue based framework is developed for the $\mathcal{H}_\infty$ norm analysis and its norm minimization of coupled systems with time-delays, which are naturally described by delay differential algebraic equations (DDAEs). For these equations $\mathcal{H}_\infty$ norms are analyzed and their sensitivity with respect to small delay perturbations is studied. Subsequently, numerical methods for the $\mathcal{H}_\infty$ norm computation and for designing controllers minimizing the $\mathcal{H}_\infty$ norm with a prescribed structure or order, based on a direct optimization approach, are briefly addressed. The effectiveness of the approach is illustrated with a software demo. The chapter concludes by pointing out the similarities with the computation and optimization of characteristic roots of DDAEs.

7.1 Introduction

In many control applications, robust controllers are desired to achieve stability and performance requirements under model uncertainties and exogenous disturbances [22]. The design requirements are usually defined in terms of $\mathcal{H}_\infty$ norms of closed-loop transfer functions including the plant, the controller and weights for uncertainties and disturbances. There are robust control methods to design the optimal $\mathcal{H}_\infty$ controller for linear finite dimensional multi-input-multi-output (MIMO) systems based on Riccati equations and linear matrix inequalities (LMIs), see e.g. [5, 7] and the references therein. The order of the controller designed by these methods is typically larger or equal to the order of the plant. This is a restrictive condition for high-

S. Gumussoy (✉)
MathWorks, Natick MA, USA
e-mail: Suat.Gumussoy@mathworks.com

W. Michiels
Department of Computer Science, KU Leuven, Leuven, Belgium
e-mail: Wim.Michiels@cs.kuleuven.be

A. Seuret et al. (eds.), *Low-Complexity Controllers for Time-Delay Systems*, Advances in Delays and Dynamics 2, DOI: 10.1007/978-3-319-05576-3_7,

order plants, since low-order controllers are desired in a practical implementation. The design of fixed-order or low-order $\mathcal{H}_\infty$ controller can be translated into a non-smooth, non-convex optimization problem. Recently fixed-order $\mathcal{H}_\infty$ controllers have been successfully designed for finite dimensional linear-time-invariant (LTI) MIMO plants using a direct optimization approach [10]. This approach allows the user to choose the controller order and tunes the parameters of the controller to minimize the $\mathcal{H}_\infty$ norm under consideration. An extension to a class of retarded time-delay systems has been described in [9].

In this work we design a fixed-order or fixed-structure $\mathcal{H}_\infty$ controller in a feedback interconnection with a time-delay system. The closed-loop system is a delay differential algebraic system and its state-space representation is written as

$$\begin{aligned} E\dot{x}(t) &= A_0x(t) + \textstyle\sum_{i=1}^{m} A_i x(t-\tau_i) + Bw(t), \\ z(t) &= Cx(t). \end{aligned} \tag{7.1}$$

The time-delays $\tau_i, i = 1, \ldots, m$ are positive real numbers and the capital letters are real-valued matrices with appropriate dimensions. The input w and output z are disturbances and signals to be minimized to achieve design requirements and some of the system matrices include the controller parameters.

The system with the closed-loop Eq. (7.1) represents all interesting cases of the feedback interconnection of a time-delay plant and a controller. The transformation of the closed-loop system to this form can be easily done by first augmenting the system equations of the plant and controller. As we shall see, this augmented system can subsequently be brought in the form (7.1) by introducing slack variables to eliminate input/output delays and direct feedthrough terms in the closed-loop equations. Hence, the resulting system of the form (7.1) is obtained directly without complicated elimination techniques that may even not be possible in the presence of time-delays.

As we shall see, the $\mathcal{H}_\infty$ norm of DDAEs may be sensitive to arbitrarily small delay changes. Since small modeling errors are inevitable in any practical design we are interested in the smallest upper bound of the $\mathcal{H}_\infty$ norm that is insensitive to small delay changes. Inspired by the concept of strong stability of neutral equations [11], this leads us to the introduction of the concept of *strong* $\mathcal{H}_\infty$ *norms* for DDAEs, Several properties of the strong $\mathcal{H}_\infty$ norm are shown and a computational formula is obtained. The theory derived can be considered as the dual of the theory of strong stability as elaborated in [11, 13, 16, 17] and the references therein.

In addition, a level set algorithm for computing strong $\mathcal{H}_\infty$ norms is presented. Level set methods rely on the property that the frequencies at which a singular value of the transfer function equals a given value (the level) can be directly obtained from the solutions of a linear eigenvalue problem with Hamiltonian symmetry (see, e.g. [1, 2, 4]), allowing a two-directional search for the global maximum. For time-delay systems this eigenvalue problem is infinite-dimensional.

Therefore, we adopt a predictor-corrector approach, where the prediction step involves a finite-dimensional approximation of the problem, and the correction serves to remove the effect of the discretization error on the numerical result. The algorithm is inspired by the algorithm for $\mathscr{H}_\infty$ computation for time-delay systems of retarded type as described in [14]. However, a main difference lies in the fact that the robustness w.r.t. small delay perturbations needs to be explicitly addressed.

The numerical algorithm for the norm computation is subsequently applied to the design of $\mathscr{H}_\infty$ controllers by a direct optimization approach. In the context of control of LTI systems it is well known that $\mathscr{H}_\infty$ norms are in general non-convex functions of the controller parameters which arise as elements of the closed-loop system matrices. They are typically even not everywhere smooth, although they are differentiable almost everywhere [10]. These properties carry over to the case of strong $\mathscr{H}_\infty$ norms of DDAEs under consideration. Therefore, special optimization methods for non-smooth, non-convex problems are required. We will use a combination of BFGS, whose favorable properties in the context of non-smooth problems have been reported in [12], bundle and gradient sampling methods, as implemented in the MATLAB code HANSO.[1] The overall algorithm only requires the evaluation of the objective function, i.e., the strong $\mathscr{H}_\infty$ norm, as well as its derivatives with respect to the controller parameters whenever it is differentiable. The computation of the derivatives is also discussed in the chapter.

The presented method is frequency domain based and builds on the eigenvalue based framework developed in [15]. Time-domain methods for the $\mathscr{H}_\infty$ control of DDAEs have been described in, e.g. [6] and the references therein, based on the construction of Lyapunov–Krasovskii functionals.

The structure of the article is as follows. In Sect. 7.2 we illustrate the generality of the system description (7.1). The concept of asymptotic transfer function of DDAEs is introduced in Sect. 7.3. The definition and properties of the strong $\mathscr{H}_\infty$ norm of DDAEs are given in Sect. 7.4. The computation of the strong $\mathscr{H}_\infty$ norm is described in Sect. 7.5. The fixed-order $\mathscr{H}_\infty$ controller design is addressed in Sect. 7.6. The concept of strong stability, fixed-order (strong) stabilization and robust stability margin optimization is summarized in Section 7.7. Section 7.8 is devoted to a software demo.

Notations The notations are as follows. The imaginary identity is j. The sets of the complex, real and natural numbers are $\mathscr{C}$, $\mathbb{R}$, $\mathbb{N}$ respectively. The sets of nonnegative and strictly positive real numbers are $\mathbb{R}^+$, $\mathbb{R}_0^+$. The matrix of full column rank whose columns span the orthogonal complement of A is shown as $A^\perp$. The zero and identity matrices are 0 and I. A rectangular matrix with dimensions $n \times m$ is $A_{n\times m}$ and when square, it is abbreviated as A_n. The ith singular value of A is $\sigma_i(A)$ such that $\sigma_1(\cdot) \geq \sigma_2(\cdot) \geq \cdots$. The short notation for $(\tau_1, \ldots, \tau_m)$ is $\boldsymbol{\tau} \in \mathbb{R}^m$. The open ball of radius $\varepsilon \in \mathbb{R}^+$ centered at $\boldsymbol{\tau} \in (\mathbb{R}^+)^m$ is defined as $\mathscr{B}(\boldsymbol{\tau}, \varepsilon) := \{\boldsymbol{\theta} \in (\mathbb{R})^m : \|\boldsymbol{\theta} - \boldsymbol{\tau}\| < \varepsilon\}$.

[1] Hybrid algorithm for nonsmooth optimization, see [20].

7.2 Motivating Examples

With some simple examples we illustrate the generality of the system description (7.1).

Example 1 Consider the feedback interconnection of the system and the controller as

$$\begin{aligned} \dot{x}(t) &= Ax(t) + B_1u(t) + B_2w(t), \\ y(t) &= Cx(t) + D_1u(t), \\ z(t) &= Fx(t), \end{aligned} \qquad \text{and} \quad u(t) = Ky(t-\tau).$$

For $\tau = 0$ it is possible to eliminate the output and controller equation, which results in the closed-loop system

$$\begin{aligned} \dot{x}(t) &= Ax(t) + B_1K(I - D_1K)^{-1}Cx(t) + B_2w(t), \\ z(t) &= Fx(t). \end{aligned} \tag{7.2}$$

This approach is for instance taken in the software package HIFOO [3]. If $\tau \neq 0$, then the elimination is not possible any more. However, if we let $X = [x^T\ u^T\ y^T]^T$ we can describe the system by the equations

$$\begin{bmatrix} I & 0 & 0 \\ 0 & 0 & 0 \\ 0 & 0 & 0 \end{bmatrix} \dot{X}(t) = \begin{bmatrix} A & B_1 & 0 \\ C & D_1 & -I \\ 0 & I & 0 \end{bmatrix} X(t) - \begin{bmatrix} 0 & 0 & 0 \\ 0 & 0 & 0 \\ 0 & 0 & K \end{bmatrix} X(t-\tau) + \begin{bmatrix} B_2 \\ 0 \\ 0 \end{bmatrix} w(t),$$

$$z(t) = \begin{bmatrix} F & 0 & 0 \end{bmatrix} X(t),$$

which are of the form (7.1). Furthermore, the dependence of the matrices of the closed-loop system on the controller parameters, K, is still linear, unlike in (7.2).

Example 2 The presence of a direct feedthrough term from w to z, as in

$$\begin{aligned} \dot{x}(t) &= Ax(t) + A_1x(t-\tau) + Bw(t), \\ z(t) &= Fx(t) + D_2w(t), \end{aligned} \tag{7.3}$$

can be avoided by introducing a slack variable. If we let $X = [x^T\ \gamma_w^T]^T$, where γ_w is the slack variable, we can bring (7.3) in the form (7.1):

$$\begin{bmatrix} I & 0 \\ 0 & 0 \end{bmatrix} \dot{X}(t) = \begin{bmatrix} A & 0 \\ 0 & -I \end{bmatrix} X(t) + \begin{bmatrix} A_1 & 0 \\ 0 & 0 \end{bmatrix} X(t-\tau) + \begin{bmatrix} B \\ I \end{bmatrix} w(t),$$

$$z(t) = [F\ D_2]\, X(t).$$

Example 3 The system

$$\begin{aligned} \dot{x}(t) &= Ax(t) + B_1 w(t) + B_2 w(t-\tau), \\ z(t) &= Cx(t), \end{aligned}$$

can also be brought in the standard form (7.1) by a slack variable. Letting $X = [x^T \gamma_w^T]^T$ we can express

$$\begin{aligned} \dot{X}(t) &= \begin{bmatrix} A & B_1 \\ 0 & -I \end{bmatrix} X(t) + \begin{bmatrix} 0 & B_2 \\ 0 & 0 \end{bmatrix} X(t-\tau) + \begin{bmatrix} 0 \\ I \end{bmatrix} w(t), \\ z(t) &= [C \;\; 0]\, X(t). \end{aligned}$$

In a similar way one can deal with delays in the output z.

Using the techniques illustrated with the above examples a broad class of interconnected systems with delays can be brought in the form (7.1), where the external inputs w and outputs z stem from the performance specifications expressed in terms of appropriately defined transfer functions.

The price to pay for the generality of the framework is the increase of the dimension of the system, n, which affects the efficiency of the numerical methods. However, this is a minor problem in most applications because the delay difference equations or algebraic constraints are related to inputs and outputs, and the number of inputs and outputs is usually much smaller than the number of state variables.

7.3 Transfer Functions

Let $\mathrm{rank}(E) = n - \nu$, with $\nu \le n$, and let the columns of matrix $U \in \mathbb{R}^{n\times\nu}$, respectively $V \in \mathbb{R}^{n\times\nu}$, be a (minimal) basis for the left, respectively right null space, that is, $U^T E = 0$, $EV = 0$.

The Eq. (7.1) can be separated into coupled delay differential and delay difference equations. When we define $\mathbf{U} = \begin{bmatrix} U^\perp & U \end{bmatrix}$, $\mathbf{V} = \begin{bmatrix} V^\perp & V \end{bmatrix}$, a pre-multiplication of (7.1) with $\mathbf{U}^T$ and the substitution $x = \mathbf{V}\, [x_1^T \; x_2^T]^T$, with $x_1(t) \in \mathbb{R}^{n-\nu}$ and $x_2(t) \in \mathbb{R}^{\nu}$, yield the coupled equations

$$\begin{aligned} E^{(11)}\dot{x}_1(t) &= \textstyle\sum_{i=0}^{m} A_i^{(11)} x_1(t-\tau_i) + \sum_{i=0}^{m} A_i^{(12)} x_2(t-\tau_i) + B_1 w(t), \\ 0 &= A_0^{(22)} x_2(t) + \textstyle\sum_{i=1}^{m} A_i^{(22)} x_2(t-\tau_i) + \sum_{i=0}^{m} A_i^{(21)} x_1(t-\tau_i) + B_2 w(t), \\ z(t) &= C_1 x_1(t) + C_2 x_2(t), \end{aligned} \tag{7.4}$$

where

$$\begin{aligned} A_i^{(11)} &= U^{\perp T} A_i V^\perp, \quad A_i^{(12)} = U^{\perp T} A_i V, \\ A_i^{(21)} &= U^T A_i V^\perp, \quad A_i^{(22)} = U^T A_i V, \quad \text{for} \quad i = 0, \ldots, m \end{aligned}$$

and

$$E^{(11)} = U^{\perp^T} E V^{\perp}, \quad B_1 = U^{\perp^T} B, \quad B_2 = U^T B, \quad C_1 = C V^{\perp}, C_2 = C.V$$

We assume two nonrestrictive conditions: matrix $U^T A_0 V$ is nonsingular and the zero solution of system (7.1), with $w \equiv 0$, is strongly exponentially stable which is a necessary assumption for $\mathcal{H}_\infty$ norm optimization. For implications of the assumptions, we refer to [8].

From (7.4) we can write the transfer function of the system (7.1) as

$$T(\lambda) := C(\lambda E - A_0 - \sum_{i=1}^{m} A_i e^{-\lambda \tau_i})^{-1} B, \tag{7.5}$$

$$= [C_1 \ C_2] \begin{bmatrix} \lambda E^{(11)} - A_{11}(\lambda) & -A_{12}(\lambda) \\ -A_{21}(\lambda) & -A_{22}(\lambda) \end{bmatrix}^{-1} \begin{bmatrix} B_1 \\ B_2 \end{bmatrix}, \tag{7.6}$$

with $A_{kl}(\lambda) = \sum_{i=0}^{m} A_i^{(kl)} e^{-\lambda \tau_i}, \ \ k, l \in \{1, 2\}$.

The *asymptotic* transfer function of the system (7.1) is defined as

$$T_a(\lambda) := -CV \left(U^T A_0 V + \sum_{i=1}^{m} U^T A_i V e^{-\lambda \tau_i} \right)^{-1} U^T B \tag{7.7}$$

$$= -C_2 A_{22}(\lambda)^{-1} B_2.$$

The terminology stems from the fact that the transfer function T and the asymptotic transfer function T_a converge to each other for high frequencies.

The $\mathcal{H}_\infty$ norm of the transfer function T of the *stable* system (7.1), is defined as

$$\|T(j\omega)\|_\infty := \sup_{\omega \in \mathbb{R}} \sigma_1 \left(T(j\omega) \right).$$

Similarly, we can define the $\mathcal{H}_\infty$ norm of T_a.

7.4 The Strong H-Infinity Norm of Time-Delay Systems

In this section we analyze continuity properties of the $\mathcal{H}_\infty$ norm of the transfer function T with respect to delay perturbations, and summarize the main results of [8], to which we refer for the proofs. The function

$$\boldsymbol{\tau} \in (\mathbb{R}_0^+)^m \mapsto \|T(j\omega, \boldsymbol{\tau})\|_\infty, \tag{7.8}$$

is, in general, not continuous, which is inherited from the behavior of the asymptotic transfer function, T_a, more precisely the function

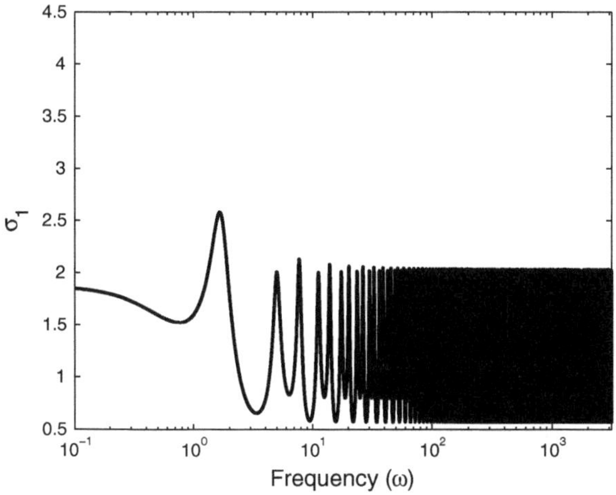

Fig. 7.1 The maximum singular value plot of $T(j\omega, \boldsymbol{\tau})$ for $(\tau_1, \tau_2) = (1, 2)$ as a function of ω

$$\boldsymbol{\tau} \in (\mathbb{R}_0^+)^m \mapsto \|T_a(j\omega, \boldsymbol{\tau})\|_\infty. \tag{7.9}$$

We start with a motivating example

Example 4 Let the transfer function T be defined as

$$T(\lambda, \boldsymbol{\tau}) = \frac{\lambda + 2.1}{(\lambda + 0.1)(1 - 0.25e^{-\lambda\tau_1} + 0.5e^{-\lambda\tau_2}) + 1} \tag{7.10}$$

where $(\tau_1, \tau_2) = (1, 2)$. The transfer function T is stable, its $\mathcal{H}_\infty$ norm is 2.5788, achieved at $\omega = 1.6555$ and the maximum singular value plot is given in Fig. 7.1. The high frequency behavior is described by the asymptotic transfer function

$$T_a(\lambda, \boldsymbol{\tau}) = \frac{1}{(1 - 0.25e^{-\lambda\tau_1} + 0.5e^{-\lambda\tau_2})}, \tag{7.11}$$

whose $\mathcal{H}_\infty$ norm is equal to 2.0320, which is less than $\|T(j\omega, \boldsymbol{\tau})\|_\infty$. However, when the first time delay is perturbed to $\tau_1 = 0.99$, the $\mathcal{H}_\infty$ norm of the transfer function T is 3.9993, reached at $\omega = 158.6569$, see Fig. 7.2. The $\mathcal{H}_\infty$ norm of T is quite different from that for $(\tau_1, \tau_2) = (1, 2)$. A closer look at the maximum singular value plot of the asymptotic transfer function T_a in Figs. 7.3 and 7.4 show that the sensitivity is due to the transfer function T_a. Even if the first delay is perturbed slightly, the problem is not resolved, indicating that the functions (7.8) and (7.9) are discontinuous at $(\tau_1, \tau_2) = (1, 2)$. When the delay perturbation tends to zero, the frequency where the maximum in the singular value plot of the asymptotic transfer function T_a is achieved moves towards infinity.

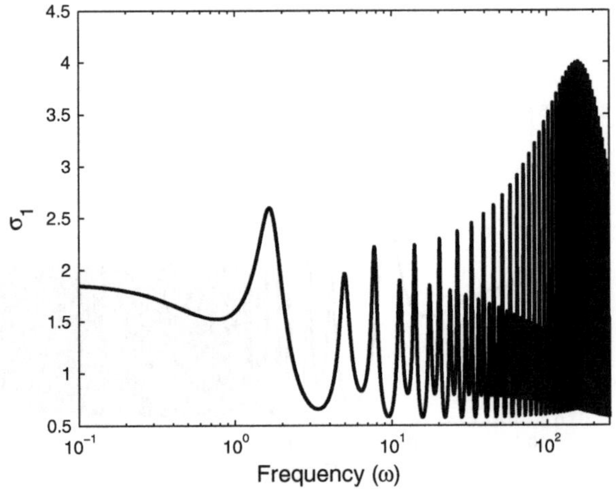

Fig. 7.2 The maximum singular value plot of $T(j\omega, \boldsymbol{\tau})$ for $(\tau_1, \tau_2) = (0.99, 2)$ as a function of ω

The above example illustrates that the $\mathscr{H}_\infty$ norm of the transfer function T may be sensitive to *infinitesimal* delay changes. On the other hand, for any $\omega_{\max} > 0$, the function

$$\boldsymbol{\tau} \mapsto \max_{[0,\ \omega_{\max}]} \sigma_1(T(jw, \boldsymbol{\tau})),$$

where the maximum is taken over a compact set, is continuous, because a discontinuity would be in contradiction with the continuity of the maximum singular value function of a matrix. Hence, the sensitivity of the $\mathscr{H}_\infty$ norm is related to the behavior of the transfer function at high frequencies and, hence, the asymptotic transfer function T_a. Accordingly we start by studying the properties of the function (7.9).

Since small modeling errors and uncertainty are inevitable in a practical design, we wish to characterize the smallest upper bound for the $\mathscr{H}_\infty$ norm of the asymptotic transfer function T_a which is *insensitive* to small delay changes.

Definition 1 For $\boldsymbol{\tau} \in (\mathbb{R}_0^+)^m$, let the strong $\mathscr{H}_\infty$ norm of T_a, $|||T_a(j\omega, \boldsymbol{\tau})|||_\infty$, be defined as

$$|||T_a(j\omega, \boldsymbol{\tau})|||_\infty := \lim_{\varepsilon \to 0+} \sup\{\|T_a(j\omega, \boldsymbol{\tau}_\varepsilon)\|_\infty : \boldsymbol{\tau}_\varepsilon \in \mathscr{B}(\boldsymbol{\tau}, \varepsilon) \cap (\mathbb{R}^+)^m\}.$$

Several properties of this upper bound on $\|T_a(j\omega, \boldsymbol{\tau})\|_\infty$ are listed below.

Proposition 1 *The following assertions hold:*

1. *For every $\boldsymbol{\tau} \in (\mathbb{R}_0^+)^m$, we have*

$$|||\ T_a(j\omega, \boldsymbol{\tau})|||_\infty = \max_{\boldsymbol{\theta} \in [0,\ 2\pi]^m} \sigma_1\left(\mathbb{T}_a(\boldsymbol{\theta})\right), \tag{7.12}$$

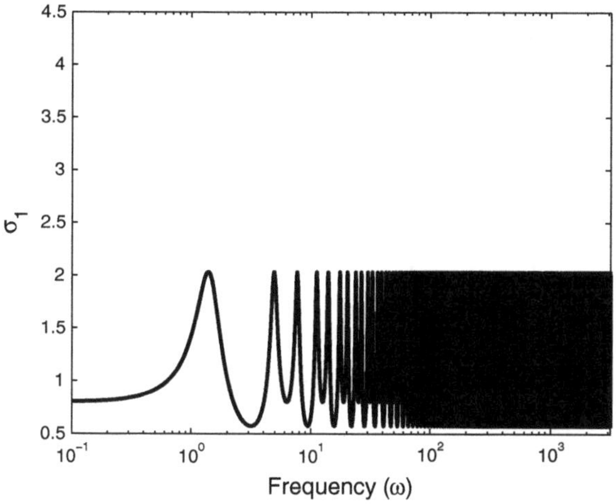

Fig. 7.3 The maximum singular value plot of $T_a(j\omega, \boldsymbol{\tau})$ for $(\tau_1, \tau_2) = (1, 2)$ as a function of ω

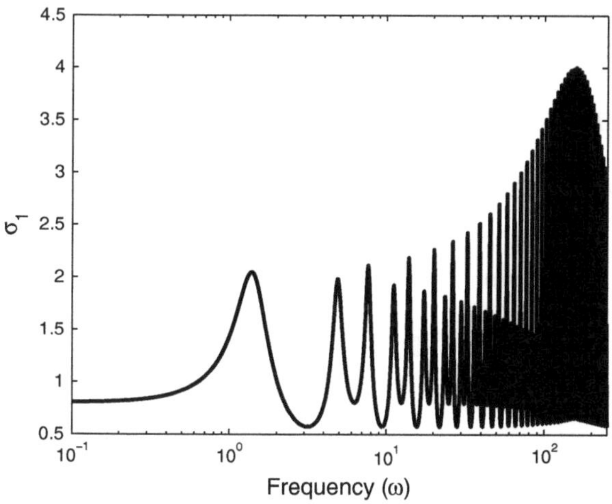

Fig. 7.4 The maximum singular value plot of $T_a(j\omega, \boldsymbol{\tau})$ for $(\tau_1, \tau_2) = (0.99, 2)$ as a function of ω

where

$$\mathbb{T}_a(\boldsymbol{\theta}) = -CV\left(U^T A_0 V + \sum_{i=1}^{m} U^T A_i V e^{-j\theta_i}\right)^{-1} U^T B; \tag{7.13}$$

2. $|||T_a(j\omega, \boldsymbol{\tau})|||_\infty \geq \|T_a(j\omega, \boldsymbol{\tau})\|_\infty$ *for all delays* $\boldsymbol{\tau}$;
3. $|||T_a(j\omega, \boldsymbol{\tau})|||_\infty = \|T_a(j\omega, \boldsymbol{\tau})\|_\infty$ *for rationally independent*[2] $\boldsymbol{\tau}$.

[2] The m components of $\boldsymbol{\tau} = (\tau_1, \ldots, \tau_m)$ are rationally independent if and only if $\sum_{k=1}^{m} z_k \tau_k = 0$, $z_k \in \mathbb{Z}$ implies $z_k = 0$, $\forall k = 1, \ldots, m$. For instance, two delays τ_1 and τ_2 are rationally independent if their ratio is an irrational number.

Formula (7.12) in Proposition 1 shows that the strong $\mathscr{H}_\infty$ norm of T_a is *independent* of the delay values. The formula further leads to a computational scheme based on sweeping on $\boldsymbol{\theta}$ intervals. This approximation can be corrected by solving a set of nonlinear equations. Numerical computation details are summarized in Sect. 7.5.

We now come back to the properties of the transfer function (7.8) of the system (7.1). As we have illustrated with Example 4, a discontinuity of the function (7.9) may carry over to the function (7.8). Therefore, we define the strong $\mathscr{H}_\infty$ norm of the transfer function T in a similar way.

Definition 2 For $\boldsymbol{\tau} \in (\mathbb{R}_0^+)^m$, the strong $\mathscr{H}_\infty$ norm of T, $|||T(j\omega, \boldsymbol{\tau})|||_\infty$, is given by

$$|||T(j\omega, \boldsymbol{\tau})|||_\infty := \lim_{\varepsilon \to 0+} \sup\{\|T(j\omega, \boldsymbol{\tau}_\varepsilon)\|_\infty : \boldsymbol{\tau}_\varepsilon \in \mathscr{B}(\boldsymbol{\tau}, \varepsilon) \cap (\mathbb{R}^+)^m\}.$$

The following main theorem describes the desirable property that, in contrast to the $\mathscr{H}_\infty$ norm, the strong H-infinity norm *continuously* depends on the delay parameters. It also presents an explicit expression that lays at the basis of the algorithm to compute the strong $\mathscr{H}_\infty$ norm of a transfer function, presented in the 7.5.

Theorem 1 *The strong $\mathscr{H}_\infty$ norm of the transfer function of the DDAE* (7.1) *satisfies*

$$|||\ T(j\omega, \boldsymbol{\tau})|||_\infty = \max\left(\|T(j\omega, \boldsymbol{\tau})\|_\infty, |||T_a(j\omega, \boldsymbol{\tau})|||_\infty\right), \tag{7.14}$$

where T and T_a are the transfer function (7.5) and the asymptotic transfer function (7.7).

In addition, the function

$$\boldsymbol{\tau} \in (\mathbb{R}_0^+)^m \mapsto |||T(j\omega, \boldsymbol{\tau})|||_\infty, \tag{7.15}$$

is continuous.

Example 5 We come back to Example 4. The $\mathscr{H}_\infty$ norm of T, as defined by (7.10), is 2.6422 and the strong $\mathscr{H}_\infty$ norm of the corresponding asymptotic transfer function T_a is 4. From property (7.14), we conclude that the strong $\mathscr{H}_\infty$ norm of T (7.10) is 4.

Remark 1 In contrast to delay perturbations, the $\mathscr{H}_\infty$ norm of T is continuous with respect to changes of the system matrices $A_i, \ldots, A_m$, B and C.

7.5 Computation of Strong H-Infinity Norms

We briefly outline the main steps of the strong $\mathscr{H}_\infty$ norm computation. Further details can be found in [8]. The algorithm for computing the strong $\mathscr{H}_\infty$ norm of the transfer function of (7.1) is based on property (7.14). This algorithm has two important steps:

1. Compute the strong $\mathscr{H}_\infty$ norm of the asymptotic transfer function T_a.
2. By taking the norm in Step 1 as the initial level set, compute the strong $\mathscr{H}_\infty$ norm of T by a level set algorithm using a predictor-corrector approach.

In the first step, the computation of $|||T_a(j\omega, \boldsymbol{\tau})|||_\infty$ is based on expression (7.12) in Proposition 1. We obtain an approximation by restricting $\boldsymbol{\theta}$ in (7.12) to a grid,

$$||| \, T_a(j\omega, \boldsymbol{\tau})|||_\infty \approx \max_{\boldsymbol{\theta} \in \Theta_h} \sigma_1 \left(\mathbb{T}_a(\boldsymbol{\theta})\right), \tag{7.16}$$

where Θ_h is a m-dimensional grid over the hypercube $[0, \; 2\pi]^m$ and $\mathbb{T}_a(\boldsymbol{\theta})$ is defined by (7.13).

If a high accuracy is required, then the approximate results may be corrected by solving a system of nonlinear equations. These equations impose that the strong $\mathscr{H}_\infty$ norm value is the maximum singular value of $\mathbb{T}_a(\boldsymbol{\theta})$, and that the derivatives of this singular value with respect to the elements of $\boldsymbol{\theta}$ are zero.

In most practical problems, the number of delays to be considered in $\mathbb{T}_a(\boldsymbol{\theta})$ is much smaller than the number of system delays, m, because most of the time-delays do not appear in $\mathbb{T}_a(\boldsymbol{\theta})$. This significantly reduces the computational cost of the sweeping in (7.16). Note that in a control application a nonzero term in (7.16) corresponds to a high frequency feedthrough over the control loop.

In the second step, the transfer function T of (7.1) is approximated by a spectral discretization. The standard level set method is applied to compute an approximation of the maximum in the singular value plot and the corresponding frequency by taking as starting level the strong $\mathscr{H}_\infty$ norm of the asymptotic transfer function T_a. For each level, a generalized eigenvalue problem is solved, from which intersections of singular value curves of the approximated system with the level set are computed. The predicted maxima and the frequencies are corrected by solving nonlinear equations characterizing a local maximum in the singular value plot of T.

7.6 Fixed-Order H-Infinity Controller Design

We consider the equations

$$\begin{array}{ll} E\dot{x}(t) &= A_0(p)x(t) + \sum_{i=1}^{m} A_i(p)x(t-\tau_i) + Bw(t), \\ z(t) &= Cx(t), \end{array} \tag{7.17}$$

where the system matrices smoothly depend on parameters p. As illustrated in Sect. 7.2, a broad class of interconnected systems can be brought into this form, where the parameters p can be interpreted in terms of a parameterization of a controller. Note that, by fixing some elements of these matrices, additional structure can be imposed on the controller, e.g. a proportional-integrative-derivative (PID) like structure.

The proposed method for designing fixed-order/ fixed-structure $\mathcal{H}_\infty$ controllers is based on a direct minimization of the strong $\mathcal{H}_\infty$ norm of the closed-loop transfer function T from w to z as a function of the parameters p. The overall optimization algorithm requires the evaluation of the objective function and its gradients with respect to the optimization parameters, whenever it is differentiable. The strong $\mathcal{H}_\infty$ norm of the transfer function T can be computed as explained in the Sect. 7.5. The computation of the derivatives of the norm with respect to controller parameters are given in [9, 19]. The overall design procedure is fully automated and does not require any interaction with the user. Further details on the design procedure can be found in [8].

7.7 Strong Stability, Fixed-Order Stabilization and Robust Stability Margin Optimization

In a practical control design, the stabilization phase is usually the first step in the overall design procedure. It is important to take the sensitivity of stability with respect to small delay perturbations into account in designing a stabilizing controller. Similarly to the $\mathcal{H}_\infty$ norm, the spectral abscissa function, i.e., the real part of the rightmost characteristic root of a system, may namely not be a continuous function of the delays [13, 17]. This implies that, although the characteristic roots of the overall system lie in the complex left half-plane, the system can become unstable when applying arbitrarily small delay perturbations. This discontinuity is due to the behavior of characteristic roots with high frequencies (imaginary parts). The counterpart of the *asymptotic transfer function* is the associated *delay difference equation* of the time delay system, and its characteristic roots with high imaginary parts correspond to these of the original system. *The robust spectral abscissa* function introduced in [18] is the smallest upper bound on the spectral abscissa which continuously depends on the delays. We say that the system is *strongly exponentially stable* if the exponential stability is robust with respect to small delay perturbations. A necessary and sufficient condition is given by a strictly negative robust spectral abscissa. An algorithm to compute the robust spectral abscissa and its derivatives with respect to controller parameters is presented in [18]. Using this algorithm and the non-smooth, non-convex optimization methods, the robust spectral abscissa is minimized and the overall system is strongly stabilized. Note that when the standard spectral abscissa function is used as objective function, the well-known fixed-order stabilization problem is solved.

Another robustness measure is the maximum value of the spectral abscissa when perturbations are considered to the system matrices whose Euclidean norm is bounded by a given constant ε. This measure is called the pseudospectral abscissa and has an interpretation in terms of a $\mathcal{H}_\infty$ norm. Inherited from this connection, the pseudospectal abscissa may also be sensitive to arbitrary small delay perturbations. In accordance, the *robust pseudospectal abscissa* can be defined, taking into account delay perturbations, in the same way as for the spectral abscissa and the $\mathscr{H}_\infty$ norm cases. Its computation is based on the computation of strong $\mathscr{H}_\infty$ norms. Using this computational method and non-smooth, non-convex optimization methods, the overall system can be stabilized under bounded perturbations on system matrices and arbitrary small perturbations on delays.

7.8 Illustration of the Software

A MATLAB implementation of the robust stabilization algorithms is available from

`http://twr.cs.kuleuven.be/research/software/delay-control/.`

Installation instructions can be found in the corresponding README file.

We consider the following system with input delay from [21]:

$$\dot{x}(t) = \mathrm{A}x(t) + w(t) + \mathrm{B}u(t-\mathrm{h}), \quad y(t) = x(t), \quad z(t) = x(t), \quad u(t) = \mathrm{k}x(t)$$

where $\mathrm{h} = 5$ and $\mathrm{k} \in \mathbb{R}^{1\times 3}$. We start by defining the system for $w \equiv 0$:

```
A = [-0.08 -0.03 0.2;0.2 -0.04 -0.005;-0.06 0.2 -0.07];
B = [-0.1;-0.2;0.1];
C = eye(3);
p1 = tds_create({A},0,{B},5,{C},0);
```

The uncontrolled system is unstable with a pole at 0.1081.

In order to compute a controller , we call a routine to minimize the robust spectral abscissa with a controller order zero, `nC=0`,

```
 [k1,f1] = stabilization_max(p1,nC);
```

The controller `k1` with the optimized robust spectral abscissa `f1` is given by:

```
 k1 =

     D11: {[0.4712 0.5037 0.6023]}
    hD11: 0

f1 =

 -0.1495
```

where empty fields of the controller are omitted for space considerations.

We inspect the characteristic roots of the closed-loop system with and without a controller by the following code. We first calculate the closed-loop with zero controller and the computed controller:

```
k0 =tds_create({},0,{},0,{},0,{[0 0 0]},0);
clp0 = closedloop(p1,k0);
clp1 = closedloop(p1,k1);
```

We can compute all eigenvalues with real part larger than -0.8 for both closed-loop systems,

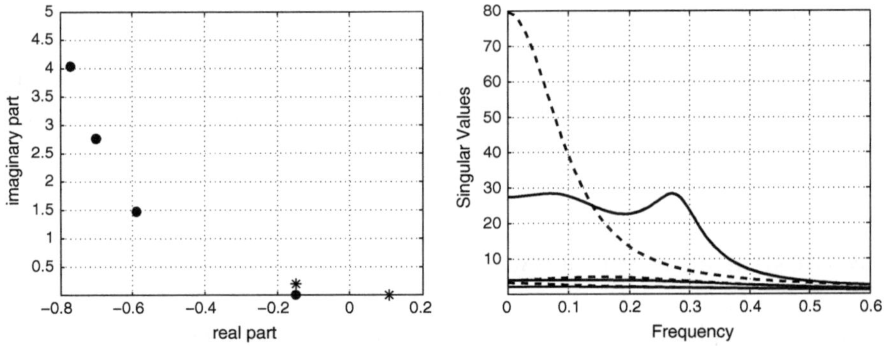

Fig. 7.5 (*left*) Characteristic roots of the open-loop system (marked with $*$) and the closed-loop system using a static controller $k1$ (marked with $\bullet$). The closed-loop system has the rightmost characteristic root at -0.1495 with a multiplicity of four. (*right*) The singular values of the closed-loop system corresponding to the minimum of the robust spectral abscissa using a static controller $k1$ (shown in dashed lines) and corresponding to the minimum of the strong $\mathscr{H}_\infty$ norm using a static controller $k2$ (shown in *straight lines*)

```
options = tdsrootsoptions;
options.minimal_real_part = -0.8;
eigenvalues0 = compute_roots_DDAE(clp0,options);
eigenvalues1 = compute_roots_DDAE(clp1,options);
```

We plot the characteristic roots of the closed-loop systems,

```
p0 = eigenvalues0.l1; plot(real(p0),imag(p0),+);
p1 = eigenvalues1.l1; plot(real(p1),imag(p1),*);
```

The results are displayed in Fig. 7.5 on the left. Note that the static controller stabilizes the closed-loop system by pushing the characteristic roots to the left of $s = -0.1495$ which corresponds to the computed robust spectral abscissa `f1` above.

In control applications, the robustness and performance objectives are often formulated as the $\mathscr{H}_\infty$ norms of transfer functions. We can tune the controller parameters of the controller K to minimize the strong $\mathscr{H}_\infty$ norm of the closed-loop system by initializing the static controller `k1` computed before,

```
% redefine plant with performance channels
p1 = tds_create({A},0,{eye(3)},0,{eye(3)},0,{},[],{B},5,{C});
% initialize the controller
options.K.initial = k1;
[k2,f2] = tds_hiopt(p1,nC,options);
```

The controller `k2` with the optimized strong $\mathscr{H}_\infty$ norm `f2` is given by:

```
 k2 =

     D11: {[0.7580 1.2247 0.6626]}
    hD11: 0


f2 =

 28.4167
```

where empty fields of the controller are omitted for space considerations.

The singular values of the closed-loop transfer function from w to z are displayed in Figure 7.5 on the right. Note that the static controller minimizing the robust spectral abscissa has a large $\mathcal{H}_\infty$ norm, 79.5443. This is expected since the controller is not tuned to minimize strong $\mathcal{H}_\infty$ norm but the robust spectral abscissa. The static controller minimizing strong $\mathcal{H}_\infty$ norm reduces the objective function to 28.4167 as indicated by `f2` and flattens the singular value plot as expected.

Acknowledgments This article present results of the Belgian Programme on Interuniversity Poles of Attraction, initiated by the Belgian State, Prime Ministers Office for Science, Technology and Culture, of the Optimization in Engineering Centre OPTEC, of the project STRT1-09/33 of the K.U.Leuven Research Council and of the Project G.0712.11 of the Fund for Scientific Research -Flanders.

References

1. Boyd, S., Balakrishnan, V., Kabamba, P.: A bisection method for computing the $\mathcal{H}_\infty$ norm of a transfer matrix and related problems. Math. Control Sig. Syst. **2**, 207–219 (1989)
2. Bruinsma, N.A., Steinbuch, M.: A fast algorithm to compute the $\mathcal{H}_\infty$-norm of a transfer function matrix. Syst. Control Lett. **14**, 287–293 (1990)
3. Burke J.V., Henrion. D., Lewis, A.S., Overton, M.L. HIFOO—a matlab package for fixed-order controller design and H-infinity optimization. In Proceedings of the 5th IFAC Symposium on Robust Control Design, Toulouse, France (2006)
4. Byers, R.: A bisection method for measuring the distance of a stable matrix to the unstable matrices. SIAM J. Sci. Stat. Comput. **9**(9), 875–881 (1988)
5. Doyle, J.C., Glover, K., Khargonekar, P.P., Francis, B.A.: State-space solutions to standard $\mathcal{H}^2$ and $\mathcal{H}^\infty$ control problems. IEEE Trans. Autom. Control **34**(8), 831–847 (1989)
6. Fridman, E., Shaked, U.: H_∞-control of linear state-delay descriptor systems: an LMI approach. Linear Algebra Appl. **351–352**, 271–302 (2002)
7. Gahinet, P., Apkarian, P.: A linear matrix inequality approach to $\mathcal{H}_\infty$ control. Int. J. Robust Nonlinear Control **4**(4), 421–448 (1994)
8. Gumussoy, S., Michiels, W.: Fixed-Order H-infinity control for interconnected systems using delay differential algebraic equations. SIAM J. Control Optim. **49**(2), 2212–2238 (2011)
9. Gumussoy, S., Michiels, W.: Fixed-order H-infinity optimization of time-delay systems. In: Diehl, M., Glineur, F., Jarlebring, E., Michiels, W. (eds.) Recent Advances in Optimization and its Applications in Engineering. Springer (2010)
10. Gumussoy, S., Overton, M.L.: Fixed-order H-infinity controller design via HIFOO, a specialized nonsmooth optimization package. In: Proceedings of the American Control Conference, pp. 2750–2754. Seattle, USA (2008)
11. Hale, J.K., Verduyn Lunel, S.M.: Strong stabilization of neutral functional differential equations. IMA J. Math. Control Inf. **19**, 5–23 (2002)
12. Lewis A and Overton M.L.: Nonsmooth optimization via BFGS. http://cs.nyu.edu/overton/papers.html (2009)
13. Michiels, W., Engelborghs, K., Roose, D., Dochain, D.: Sensitivity to infinitesimal delays in neutral equations. SIAM J. Control Optim. **40**(4), 1134–1158 (2002)
14. Michiels, W., Gumussoy, S.: Characterization and computation of H-infinity norms of time-delay systems. J. Matrix Anal. Appl. **31**(4), 2093–2115 (2010)
15. Michiels, W., Niculescu S.-I.: Stability and stabilization of time-delay systems. An eigenvalue based approach. SIAM (2007)

16. Michiels, W., Vyhlídal, T.: An eigenvalue based approach for the stabilization of linear time-delay systems of neutral type. Automatica **41**(6), 991–998 (2005)
17. Michiels, W., Vyhlídal, T., Zítek, P., Nijmeijer, H., Henrion, D.: Strong stability of neutral equations with an arbitrary delay dependency structure. SIAM J. Control Optim. **48**(2), 763–786 (2009)
18. Michiels, W., Gumussoy, S.: Eigenvalue based algorithms and software for the design of fixed-order stabilizing controllers for interconnected systems withtime-delays. Delay Systems, Advances in Delays and Dynamics, pp.243–256. Springer (2013)
19. Millstone, M.: HIFOO 1.5: Structured control of linear systems with a non-trivial feedthrough. Master's thesis, New York University (2006)
20. Overton, M.: HANSO: a hybrid algorithm for nonsmooth optimization. http://cs.nyu.edu/overton/software/hanso/ (2009)
21. Vanbiervliet, J., Vandereycken, B., Michiels, W., Vandewalle, S.: A nonsmooth optimization approach for the stabilization of time-delay systems. ESAIM Control Optim. Calc. Var. **14**(3), 478–493 (2008)
22. Zhou, K., Doyle, J.C., Glover, K.: Robust and optimal control. Prentice Hall, Englewood Cliffs, NJ (1995)

Chapter 8
Delay-Independent Stability Via Reset Loops

Alfonso Baños, Félix Pérez Rubio, Sophie Tarbouriech and Luca Zaccarian

Abstract In this chapter we propose a control scheme where a pre-designed linear controller in feedback with a time-delay plant is augmented with suitable jump rules that are activated in certain subsets of the state space to ensure closed-loop asymptotic stability. Under suitable feasibility conditions on the data of the linear time-delay plant, we show that the proposed scheme induces delay-independent stability of the closed loop with controller state jumps. Due to the hybrid nature of the proposed scheme, we address stability by proposing a hybrid version of the classical Lyapunov-Krasovskii theorem, relying on a dwell-time condition and on a Lyapunov-Krasovskii function that does not increase accross jumps. The results in the chapter can be seen as preliminary results in the direction of hybrid time-delay dynamical systems, which still remains largely unexplored. A simulation example shows the effectiveness of the proposed hybrid scheme.

A. Baños (✉)
Departamento de Informatica y Sistemas, Facultad de Informatica,
Universidad de Murcia, Murcia, Spain
e-mail: abanos@um.es

F. Pérez Rubio
Software Department, MTorres SAU, 31119 Navarra, Spain
e-mail: Felix.Perez@mtorres.es

S. Tarbouriech · L. Zaccarian
CNRS, LAAS, 7 avenue du colonel Roche, 31400 Toulouse, France
e-mail: tarbour@laas.fr

S. Tarbouriech · L. Zaccarian
Univ de Toulouse, LAAS, 31400 Toulouse, France
e-mail: zaccarian@laas.fr

L. Zaccarian
Dipartimento di Ingegneria Industriale, University of Trento, Trento, Italy
e-mail: zaccarian@laas.fr

A. Seuret et al. (eds.), *Low-Complexity Controllers for Time-Delay Systems*,
Advances in Delays and Dynamics 2, DOI: 10.1007/978-3-319-05576-3_8,

8.1 Introduction

An important aspect that must be taken into account in many control applications is the presence of delay in the loop, which can be the source of performance degradation or even instability (see, for example, [17, 18] and references therein).

A control architecture that received much attention in recent years is that of reset control. Reset control systems can be represented by means of different hybrid systems formulations. Many of the previous work about systems with resets has been done by using the impulsive state dependent state formulation of [10]. Alternatively, the hybrid formulation developed in [7, 8] has been also used for describing reset control systems, and formal stability results have been already obtained (see, e.g., [14, 15, 19]). This last framework has also been used for systems with time delays in the recent work [13], that is inspiring for the notation followed in this chapter. This formulation will be first used for stability analysis of reset control system with time delays, and with a new characterization of flow and jump sets, inspired by [6], that allows us to state a Lyapunov-Krasovskii theorem for delay-independent stability of the reset control system.

From a historical perspective, the idea of reset compensation dates back to the seminal works of Clegg and Horowitz [5, 12], where simple reset compensators such as the Clegg integrator or the First Order Reset Element (FORE) were first proposed with the main motivation of overcoming fundamental limitations of linear time-invariant compensators in control practice. Intuitively, within this quest, since it was observed already in [5] (see also [2]) that reset compensation introduces phase lead without significantly increasing the loop gain, it seems to be appropriate to propose reset compensation for systems with time-delays due to the evident time lag introduced by the delay effect. In [1], a delay-independent stability analysis of reset control system is first performed, that has been extended to the more general delay-dependent case in [2, 4] and in [9].

In this chapter, we are interested in studying the stability of closed loops independently of the delay. In particular, our work is focused on the design of hybrid reset rules based on Lyapunov conditions and applied to time-delay continuous-time plants in order to eliminate possible instability arising from the linear interconnection. We propose a kind of hybrid controller that can guarantee delay-independent stability of the closed loop. The class of systems that we address combines two ingredients: (1) the continuous-time dynamics of a linear time-delay plant and a linear controller, enforced when the overall state belongs to a certain *flow set* $\mathscr{F}$ and (2) a discrete dynamics corresponding to an impulsive action performed on the controller state when the overall state belongs to a certain *jump set* $\mathscr{J}$ [8]. It is important to emphasize that few works are dedicated to the study of hybrid time-delay systems. Notable exceptions comprise the reset systems work in [1, 9] and the framework recently proposed in [13].

The scheme proposed in this chapter arises from adapting the hybrid augmentation paradigm recently proposed in [6, 16]. Moreover, to prove our main result, we also formulate a Lyapunov-Krasovskii theorem that complements the results issued from

[1] and [9], by using a different definition of flow and jumps sets and an additional dwell-time logic.

In the sequel we first introduce the problem under consideration in Sect. 8.2. Then we describe the proposed closed loop with resets in Sect. 8.3. Subsequently we prove suitable stability properties of the control scheme in Sect. 8.4 and finally we discuss a simulation example in Sect. 8.5.

8.2 Problem Data and Standing Assumption

Consider the following strictly proper linear time-delay plant:

$$\mathscr{P}: \quad \begin{aligned} \dot{x}_p(t) &= A_p x_p(t) + A_{pd} x_p(t-\theta) + B_p u_c(t), \\ y_p(t) &= C_p x_p(t), \end{aligned} \tag{8.1}$$

where $x_p \in \mathbb{R}^{n_p}$ is the state of the plant, $\theta \in \mathbb{R}_{\geq 0}$ is a known state delay, $y_p \in \mathbb{R}^{n_y}$ is the output available for measurement and u_c is the control input to be used in the controller design. It is costumary to denote the infinite dimensional state of plant (8.3) as

$$x_{pt} = \{x_p(s), s \in [t-\theta, t]\}, \tag{8.2}$$

and to use the shortcut notation $x_{pd}(t) = x_p(t-\theta)$ so that Eq. (8.3) can be written in compact from as

$$\mathscr{P}: \quad \begin{aligned} \dot{x}_p &= A_p x_p + A_{pd} x_{pd} + B_p u_p, \\ y_p &= C_p x_p. \end{aligned} \tag{8.3}$$

For plant (8.3) we assume that a linear time invariant controller has been designed to ensure suitable closed-loop properties under certain operating conditions:

$$\mathscr{C}: \quad \begin{aligned} \dot{x}_c &= A_c x_p + B_c y_p, \\ u_p &= C_c x_c, \end{aligned} \tag{8.4}$$

where $x_c \in \mathbb{R}^{n_c}$ is the state of the controller. Controller (8.4) has been designed for plant (8.3) disregarding the effect of delay, namely in such a way to stabilize the delay-free dynamics corresponding to the transition matrix $A_p + A_{pd}$, which corresponds to plant (8.3) in the special case $\theta = 0$. Then the goal of this paper is to introduce suitable reset rules on the controller states that ensure recovery of closed-loop asymptotic stability for any known value of the delay θ. Since we do not need the requirement that controller (8.4) stabilizes the plant when $\theta = 0$, we don't make this as an explicit assumption.

In this work we will use Lyapunov-Krasovskii techniques to assess asymptotic stability of the closed loop for any value of the time delay θ, that is, stability independent of the delay. Due to this fact, we will require the following assumption on the plant data.

Assumption 1 *Given the matrices in (8.3) there exist two positive definite matrices* P_p *and* Q*, a gain* K_p *and a scalar* $\varepsilon_p > 0$ *such that*

$$He \begin{bmatrix} P_p(A_p + B_p C_c K_p) + Q/2 & P_p A_{pd} \\ 0 & -Q/2 \end{bmatrix} \leq -2\varepsilon_p \begin{bmatrix} P_p & 0 \\ 0 & Q \end{bmatrix}. \tag{8.5}$$

Assumption 1 ensures that it is possible to prestabilize plant (8.3) by way of the state feedback gain $u_p = K_p x_p$ and obtain a delay-independent stable closed loop. The following lemma clarifies that the search for parameters P_p, Q and K_p in (8.5) is equivalent to a convex (LMI eigenvalue) problem.

Lemma 1 *Assumption 1 holds if and only if the following LMI in the variables* $Q_p = Q_p^T > 0$*,* $S = S^T > 0$ *and* X *is feasible:*

$$He \begin{bmatrix} A_p Q_p + B_p C_c X + S/2 & A_{pd} Q_p \\ 0 & -S/2 \end{bmatrix} < 0. \tag{8.6}$$

Moreover, whenever (8.6) holds, a solution to Assumption 1 is given by $P_p = Q_p^{-1}$*,* $Q = P_p S P_p$*,* $K_p = X P_p$ *and a small enough* ε_p*.*

Proof Consider Eq. (8.6) and perform the congruence transformation pre- and post-multiplying by the block-diagonal symmetric matrix $\mathrm{diag}(Q_p^{-1}, S^{-1})$. Then the following equivalent relation to (8.6) is obtained, with the definitions in the lemma:

$$\mathrm{He} \begin{bmatrix} P_p(A_p + B_p C_c K_p) + Q/2 & P_p A_{pd} \\ 0 & -Q/2 \end{bmatrix} < 0. \tag{8.7}$$

Assume now that Assumption 1 holds, then obviously Eq. (8.7) holds too, which is equivalent to (8.6). Viceversa, if (8.6) (therefore (8.7)) holds, then due to the strict inequality in (8.7) there exists a small enough $\varepsilon_p > 0$ such that (8.5) holds with the selections in the statement of the lemma.

Remark 1 Once the feasibility condition (8.6) is verified, it might be of interest to seek for the solution to (8.5) corresponding to maximizing ε_p while imposing that the gain K_p satisfies a prescribed bound $|K_p| \leq \kappa_M$, for some fixed scalar $\kappa_M > 0$. This solution can be computed by solving the following optimization problem:

$$\begin{aligned} \varepsilon_p^* = & \max_{Q_p, X, S, \varepsilon_p} \varepsilon_p \text{ subject to:} \\ & Q_p \geq I, \begin{bmatrix} \kappa_M I & X \\ X^T & \kappa_M I \end{bmatrix} \geq 0, \\ & \mathrm{He} \begin{bmatrix} A_p Q_p + B_p C_c X + S/2 & A_{pd} Q_p \\ 0 & -S/2 \end{bmatrix} \leq -2\varepsilon_p \begin{bmatrix} Q_p & 0 \\ 0 & S \end{bmatrix}, \end{aligned} \tag{8.8}$$

which is a generalized eigenvalue problem (namely a quasi convex optimization problem) for which efficient numerical solution algorithms are available. The corresponding solution to (8.5) can then be computed as $P_p = Q_p^{-1}$, $Q = P_p S P_p$ and $K_p = X P_p$, just as in Lemma 1.

8.3 Hybrid Closed-Loop System

In this section we design a hybrid closed-loop system whose flow dynamics corresponds to the interconnection between (8.3) and (8.4) and whose jump dynamics and jump and flow sets are constructed, based on a solution to (8.5) in Assumption 1, in such a way to guarantee uniform global asymptotic stability of the origin of the plant-controller state space. Note that this property is non-trivial because no assumption is made on the stability properties of the continuous-time interconnection (8.3), (8.4)

Adopting the notation in [7, 8], we propose the following dwell-time hybrid augmentation of the closed loop (8.3), (8.4), where for convenience of notation we denote the aggregated (and transformed) state $\xi = (x_p, x_{pd}, \delta) := (x_p, x_{pd}, x_c - K_p x_p)$:

$$\begin{cases} \dot{x}_p = A_p x_p + A_{pd} x_{pd} + B_p C_c x_c \\ \dot{x}_c = A_c x_c + B_c C_p x_p & (\xi, \tau) \in \mathscr{C} \times [0, 2\rho], \\ \dot{\tau} = 1 - \text{dz}\left(\frac{\tau}{\rho}\right) \end{cases}$$
$$\begin{cases} x_p^+ = x_p \\ x_c^+ = K_p x_p & (\xi, \tau) \in \mathscr{D} \times [\rho, 2\rho], \\ \tau^+ = 0 \end{cases} \tag{8.9a}$$

where $\text{dz}(\cdot)$ denotes the scalar unit deadzone function, the sets $\mathscr{C}$ and $\mathscr{D}$ are defined, based on an arbitrary positive scalar $\varepsilon < \varepsilon_p$, as follows:

$$\mathscr{C} := \left\{ \xi : \xi^T \text{He} \begin{bmatrix} P_p A_R + Q/2 & P_p A_{pd} & P_p B_p C_c \\ 0 & -Q/2 & 0 \\ 0 & 0 & 0 \end{bmatrix} \xi \leq -\varepsilon \xi^T \begin{bmatrix} P_p & 0 & 0 \\ 0 & Q & 0 \\ 0 & 0 & I \end{bmatrix} \xi \right\}, \tag{8.9b}$$

$$\mathscr{D} := \left\{ \xi : \xi^T \text{He} \begin{bmatrix} P_p A_R + Q/2 & P_p A_{pd} & P_p B_p C_c \\ 0 & -Q/2 & 0 \\ 0 & 0 & 0 \end{bmatrix} \xi \geq -\varepsilon \xi^T \begin{bmatrix} P_p & 0 & 0 \\ 0 & Q & 0 \\ 0 & 0 & I \end{bmatrix} \xi \right\}. \tag{8.9c}$$

In the sets $\mathscr{C}$ and $\mathscr{D}$, A_R is defined as $A_R := A_p + B_p C_c K_p$ and matrices P_p and Q are defined in Assumption 1.

Following [8] and inspired by the definitions in [13], we introduce the following definitions to suitably characterize solutions to hybrid system (8.9a–8.9c). In particular, the following definitions are slightly different from those in [13] because we are exploiting here the property that the memory of the time-delay system is only in the directions of the plant state, which remains unchanged across jumps.

Definition 1 A subset $E \subset \mathbb{R}_{\geq 0} \times \mathbb{Z}_{\geq 0}$ is called a *compact hybrid time domain* if

$$E = \bigcup_{j=0}^{J-1} ([t_j, t_{j+1}], j),$$

for some finite sequences of times $0 = t_0 \leq t_1 \leq \cdots \leq t_J$, called "jump times". The set E is called a *hybrid time domain* if for all $(T, J) \in E$, the set $E \cap ([0, T], \{0, 1, \ldots, J\})$ is a compact hybrid time domain.

Given a positive real $\theta > 0$, a subset $E_0 \subset \mathbb{R}_{\geq -\theta} \times \mathbb{Z}_{\leq 0}$ is called a *hybrid time domain with ordinary memory* θ if

$$E_0 = ([-\theta, 0), 0) \cup E,$$

where E is a hybrid time domain.

A *hybrid arc with ordinary plant memory* θ is a triple consisting of a domain $\mathrm{dom}\phi$ that is a hybrid time domain with ordinary memory θ, a continuous function $\phi_{p0}(s)$, $s \in [-\theta, 0]$ representing the (infinite dimensional) initial condition of the system in the plant state direction and a function $\phi : \mathrm{dom}_{\geq 0}\phi \to \mathbb{R}^{n_p \times n_c}$, where $\mathrm{dom}_{\geq 0}\phi := \mathrm{dom}\phi \cap (\mathbb{R}_{\geq 0} \times \mathbb{Z}_{\geq 0})$, such that $\phi(\cdot, j)$ is locally absolutely continuous on $I^j = \{t : (t, j) \in \mathrm{dom}_{\geq 0}\phi\}$.

Similar to [13, Def. 4], but for the special case where the memory of the time-delay system is only in the ordinary time direction, given any hybrid arc $\phi = (\phi_p, \phi_c)$ with ordinary plant memory θ, for each $(t, j) \in \mathrm{dom}_{\geq 0}\phi$, we define the operator $\mu_{[t,j]}\phi_p(s)$ that maps the hybrid arc ϕ into an ordinary memory arc of length θ (this is a function of ordinary time only). In particular, $\mathrm{dom}\mu_{[t,j]}\phi_p = [-\theta, 0]$ and for each $s \in [-\theta, 0]$ we have $\mu_{[t,j]}\phi_p(s) = \phi_p(t + s, i)$ for some $i \in \mathbb{Z}_{\geq 0}$ such that $(t - s, i) \in \mathrm{dom}\phi$. Note that such an i exists because the hybrid arc ϕ has ordinary plant memory θ by assumption. Note also that the definition above has no ambiguity because of the special structure in (8.9a–8.9c) where the p component of the solution remains constant across jumps. Indeed, if there exist multiple values $i_1, i_2, \in \mathbb{Z}_{\geq 0}$ such that $(t - s, i_1), (t - s, i_2) \in \mathrm{dom}\phi$, then we have $\mu_{[t,j]}\phi_p(s) = \phi_p(t + s, i_1) = \phi_p(t + s, i_2)$. Indeed, projecting the memory of the time-delay system only on the ordinary time domain axis (by way of the operator $\mu_{[t,j]}$) greatly simplifies the forthcoming derivations.

Based on the above characterization, we can formulate a class of hybrid time-delay systems that generalizes the peculiar structure of (8.9a–8.9c) and that can be written in compact form as follows:

$$\begin{aligned}
\begin{bmatrix} \dot{x}_p \\ \dot{x}_c \\ \dot{\tau} \end{bmatrix} &= \begin{bmatrix} f_p(x_p, x_c, \mu x_p) \\ f_c(x_p, x_c, \mu x_p) \\ 1 - \mathrm{dz}\left(\frac{\tau}{\rho}\right) \end{bmatrix}, \quad \begin{bmatrix} x_p \\ x_c \\ \mu x_p \\ \tau \end{bmatrix} \in \widetilde{\mathscr{C}} \times [0, 2\rho], \quad (x_p(0,0), x_c(0,0)) \in \mathbb{R}^{n_p \times n_c} \\
\begin{bmatrix} x_p^+ \\ x_c^+ \\ \tau^+ \end{bmatrix} &= \begin{bmatrix} x_p \\ g_c(x_p, x_c, \mu x_p) \\ 0 \end{bmatrix}, \quad \begin{bmatrix} x_p \\ x_c \\ \mu x_p \\ \tau \end{bmatrix} \in \widetilde{\mathscr{D}} \times [\rho, 2\rho], \quad \mu_{[0,0]} x_p \in \mathscr{C}^0([-\theta, 0]),
\end{aligned} \tag{8.10}$$

where $\mathscr{C}^0(\mathscr{T})$ denotes the set of continuous functions whose domain is $\mathscr{T} \subset \mathbb{R}$, f_p, f_c and g_c are suitable linear functionals that, for system (8.9a–8.9c), correspond to:

$$\begin{aligned}
f_p(x_p, x_c, \mu x_p) &= A_p x_p + A_{pd} \mu x_p(-\theta) + B_p C_c x_c, \\
f_c(x_p, x_c, \mu x_p) &= A_c x_c + B_c C_p x_p, \\
g_c(x_p, x_c, \mu x_p) &= K_p x_p.
\end{aligned}$$

Moreover, the flow and jump sets $\widetilde{\mathscr{C}}$ and $\widetilde{\mathscr{D}}$ are suitable infinite dimensional flow and jump sets.

Based on Definition 1, a solution to hybrid time-delay system (8.10) is a hybrid arc with ordinary plant memory θ satisfying the flow and jump constraints imposed by the hybrid dynamics. A precise formulation of this can be obtained by following the same paradigm as that used in [13, Def. 6].

Remark 2 The extension of the formalism in [7, 8] to the time-delay framework is currently underdeveloped (a notable exception being the recent paper [13]). Therefore, not much can be stated about existence of solutions or even nominal or robust well posedness of solutions to these hybrid systems in the sense of [8, Chap. 6]. Results on well posedness of a particular class of hybrid systems is given in [13], however system (8.10) goes beyond this class because in [13] it is assumed that the flow and jump sets are subsets of the Euclidean spaces where x_p and x_c take values. Here we have a new direction in the jump and flow sets depending on the (infinite dimensional) memory of the plant state x_p and it is not evident how to inherit the results of [13]. However, at least from the existence viewpoint, the architecture (8.10) can still inherit useful properties from the classical results in [11] and from the results in [8, Chap. 6], as long as the initial condition $\mu_{[0,0]} x_p$ of the plant is continuous.

Remark 3 Note that differently from [6], the dwell-time logic is implemented in (8.9a–8.9c) by forcing solutions not to jump unless $\tau \geq \rho$, while also not allowing them to flow if their ξ component does not belong to the set $\mathscr{C}$. Because of this fact, dwell time is artificially enforced on solutions by possibly terminating defective solutions that would jump too often. The natural question that arises is then whether system (8.9a–8.9c) admits complete solutions for all initial conditions starting either in the jump or in the flow set. This question is not addressed here, where we simply limit ourselves to observing that the example treated in Sect. 8.5 exhibits complete solutions. We regard tackling this important aspect as future work.

8.4 Stability Properties of the Reset Control Scheme

8.4.1 A Lyapunov-Krasovskii Theorem

We state in this section a peculiar version of a Lyapunov-Krasovskii theorem for system (8.10). Alternative instances of this type of result have appeared in the recent literature. For example see [1, Prop. 3.1] or [9, Prop. 1]. We state here a different formulation of the result, due to the special definitions and hybrid framework used in the previous section.

Since we are not interested in the evolution of the timer τ within the compact set $[0, 2\rho]$ where it is confined, we will characterize stability properties of the following compact attractor for dynamics (8.10):

$$\mathscr{A} := \{0\} \times [0, 2\rho] \subset \mathbb{R}^{n_p+n_c} \times \mathbb{R}. \tag{8.11}$$

In particular, following standard derivations in the time-delay continuous-time systems framework, given any solution $\phi = (\phi_x, \phi_\tau) = (\phi_p, \phi_c, \phi_\tau)$ to (8.10), we introduce the following notion of distance from the set $\mathscr{A}$ in (8.11) for each $(t, j) \in \text{dom}_{\geq 0}\phi$:

$$\|\phi(t, j)\|_\mu = \max\left\{|\phi_x(t, j)|, \max_{s\in[-\theta,0]} |\mu_{[t,j]}\phi_p(s)|\right\}, \tag{8.12}$$

where $|\cdot|$ denotes the Euclidean norm.

Since we are dealing with a special class of hybrid time-delay systems, we clarify the meaning of stability in the following definition, which is inspired by [13] and [8, Sect. 3.1].

Definition 2 The compact set $\mathscr{A}$ in (8.11) is

1. Globally stable (GS) for (8.10) if there exists a class $\mathscr{K}_\infty$ function α such that any solution $\phi = (\phi_x, \phi_\tau) = (\phi_p, \phi_c, \phi_\tau)$ to (8.10) satisfies $|\phi_x(t, j)| \leq \alpha(\|\phi(0, 0)\|_\mu)$ for all $(t, j) \in \text{dom}_{\geq 0}\phi$.
2. Uniformly globally attractive (UGA) for (8.10) if for each pair r, ε, there exists $T(r, \varepsilon)$ such that any solution ϕ satisfies:

$$\|\phi(0, 0)\|_\mu \leq r \Rightarrow |\phi_x(t, j)| \leq \varepsilon, \ \forall(t, j) \in \text{dom}_{\geq 0}\phi \text{ such that } t+j \geq T(r, \varepsilon). \tag{8.13}$$

3. Uniformly globally asymptotically stable for (8.10) if it is GS and UGA.

Based on Definition 1 we can state a Lyapunov-Krasovskii result that only requires the following mild uniform boundedness assumption on the functions appearing in the flow map of (8.10). This assumption is trivially satisfied by the linear flow dynamics in (8.9a–8.9c).

Assumption 2 *There exists a class $\mathcal{K}_\infty$ function α_M such that for each $r \geq 0$ (where we use $x = (x_p, x_c)$),*

$$\sup_{(x,\mu x_p)\in\mathcal{C}\ s.t.\ \|x\|_\mu \leq r} \begin{bmatrix} f_p(x, \mu x_p) \\ f_c(x, \mu x_p) \end{bmatrix} \leq \alpha_M(r).$$

Proposition 1 *Under Assumption 2, if there exist a function V, two class $\mathcal{K}_\infty$ functions α_1, α_2 and a positive definite function σ satisfying*

$$\alpha_1(|x|) \leq V((x,\tau), \mu x_p) \leq \alpha_2(\|(x,\tau)\|_\mu), \quad \forall (x, \mu x_p, \tau), \tag{8.14}$$

$$\dot{V}((x,\tau), \mu x_p) \leq -\rho(|x|), \quad \forall (x, \mu x_p, \tau) \in \widetilde{\mathcal{C}} \times [0, 2\rho], \tag{8.15}$$

$$V((x^+,\tau^+), \mu x_p) - V((x,\tau), \mu x_p) \leq 0, \quad \forall (x, \mu x_p, \tau) \in \widetilde{\mathcal{D}} \times [\rho, 2\rho], \tag{8.16}$$

then the compact attractor $\mathcal{A}$ in (8.11) is uniformly globally asymptotically stable for (8.10).

Proof The proof is omitted due to space constraints but can be found in [3]. □

8.4.2 Main Stability Result

We state next our main stability result for the hybrid dynamics (8.9a–8.9c). To properly state the stability result,

Theorem 1 *Consider a plant (8.3) satisfying Assumption 1 and a controller (8.4). Then for the dwell-time hybrid time-delay dynamics (8.9a–8.9c) there exist a functional V and class $\mathcal{K}_\infty$ functions α_1, α_2 and σ satisfying (8.14)–(8.16). Namely, the set $\mathcal{A}$ in (8.11) is globally asymptotically stable.*

Proof Using P_p and Q of Assumption 1, consider the following Lyapunov-Krasovskii functional for the plant state directions:

$$V_p(x_p, \mu x_p) = x_p^T P_p x_p + \int_{-\theta}^{0} \mu x_p^T(s) Q \mu x_p(s) ds. \tag{8.17}$$

The derivative of V_p along the flow dynamics of (8.9a–8.9c) is (for notational compactness, we use $x_{pd} := \mu x_p(-\theta)$ in the rest of the proof):

$$\begin{aligned}
\dot{V}_p &= 2x_p^T P_p \left[A_p x_p + B_p C_c \left(K_p x_p + x_c - K_p x_p\right) + A_{pd} x_{pd}\right] + x_p^T Q x_p - x_{pd}^T Q x_{pd} \\
&= 2x_p^T P_p \left[\underbrace{(A_p + B_p C_c K_p)}_{A_R} x_p + A_{pd} x_{pd} + B_p C_c \underbrace{(x_c - K_p x_p)}_{\delta} \right] + x_p^T Q x_p - x_{pd}^T Q x_{pd} \\
&= \begin{bmatrix} x_p \\ x_{pd} \\ \delta \end{bmatrix}^T \mathrm{He} \begin{bmatrix} P_p A_R + Q/2 & P_p A_{pd} & P_p B_p C_c \\ 0 & -Q/2 & 0 \\ 0 & 0 & 0 \end{bmatrix} \begin{bmatrix} x_p \\ x_{pd} \\ \delta \end{bmatrix}.
\end{aligned} \tag{8.18}$$

Then, from the definition (8.9b) of the flow set $\mathscr{C}$ we have that

$$\dot{V}_p \le -\varepsilon \left(|x_p|_{P_p}^2 + |x_{pd}|_Q^2 + |\delta|^2 \right), \quad \forall (x_p, x_{pd}, \delta) \in \mathscr{C} \tag{8.19}$$

where $|y|_W = \sqrt{y^T W y}$.

Now consider the following Lyapunov-Krasovskii functional for the overall system:

$$V = V_p(x_p, \mu x_p) + \lambda \delta^2, \tag{8.20}$$

where λ is a positive scalar selected later. We show next that this function satisfies (8.14)–(8.16), so that the result follows from Proposition 1.

Proof of (8.14) The lower bound on the left comes from

$$x_p^T P_p x_p + \lambda \delta^2 = \begin{bmatrix} x_p \\ x_c \end{bmatrix}^T \underbrace{\begin{bmatrix} P_p + \lambda K_p^T K_p & -\lambda K_p \\ -\lambda K_p^T & \lambda I \end{bmatrix}}_{\Sigma} \begin{bmatrix} x_p \\ x_c \end{bmatrix} \ge \sigma_m(\Sigma) \left| \begin{bmatrix} x_p \\ x_c \end{bmatrix} \right|^2 =: \alpha_1 \left(\left| \left[\begin{smallmatrix} x_p \\ x_c \end{smallmatrix} \right] \right| \right),$$

where $\sigma_m(\Sigma) > 0$ because Σ is positive definite.

For the upper bound we get:

$$\begin{aligned}
V(x, \mu x_p) &\le \left[\begin{smallmatrix} x_p \\ x_c \end{smallmatrix} \right]^T \Sigma \left[\begin{smallmatrix} x_p \\ x_c \end{smallmatrix} \right] + \int_{-\theta}^{0} \sigma_M(Q) |\mu x_p(s)|^2 ds \\
&\le \sigma_M(\Sigma) \left| \left[\begin{smallmatrix} x_p \\ x_c \end{smallmatrix} \right] \right|^2 + \theta \sigma_M(Q) \max_{s \in [-\theta, 0]} |\mu x_p(s)|^2 \\
&\le 2 \max\{\sigma_M(\Sigma),\ \theta \sigma_M(Q)\} \|(x, \tau)\|_\mu^2 =: \alpha_2(\|(x, \tau)\|_\mu).
\end{aligned}$$

where $\sigma_M(\cdot)$ and $\sigma_m(\cdot)$ denote, respectively, the maximum and minimum singular values of their arguments.

Proof of (8.15) First, we may easily compute:

$$\dot{V} \le \dot{V}_p + \lambda \delta (\underbrace{(A_c - K_p B_p C_c)}_{A_1} x_c + \underbrace{(B_c C_p - K_p A_p)}_{A_2} x_p + \underbrace{(-K_p A_{pd})}_{A_3} x_{pd}). \tag{8.21}$$

Since $x_c = \delta + K_p x_p$, then using (8.19), Eq. (8.21) can be rewritten as

$$\begin{aligned}\dot{V} &\leq -\varepsilon|x_p|^2_{P_p} - \varepsilon|x_{pd}|^2_Q - \varepsilon|\delta|^2 + \lambda\delta(A_1\delta + (A_1K_p + A_2)x_p + A_3x_{pd}) \\ &\leq -\varepsilon\underline{c}|x_p|^2 - \varepsilon\underline{c}|x_{pd}|^2 - \varepsilon|\delta|^2 + \lambda\overline{c}(|\delta|^2 + |\delta|\,|x_p| + |\delta|\,|x_{pd}|),\end{aligned}$$

where $\overline{c} = \max\{\sigma_M(A_1), \sigma_M(A_1K_p+A_2), \sigma_M(A_3)\}$ and $\underline{c} = \min\{\sigma_m(P_p), \sigma_m(Q)\}$.
Finally, completing squares and choosing $\lambda = \varepsilon \min\left\{\frac{\underline{c}}{\overline{c}}, \frac{1}{4\overline{c}}\right\}$, we get

$$\begin{aligned}\dot{V} &\leq -\left(\varepsilon\underline{c} - \frac{\lambda\overline{c}}{2}\right)|x_p|^2 - \left(-\varepsilon\underline{c} + \frac{\lambda\overline{c}}{2}\right)|x_{pd}|^2 - (\varepsilon - 2\lambda\overline{c})|\delta|^2 \\ &\leq -\frac{\varepsilon}{2}\left|\begin{bmatrix} I & 0 \\ -K_p & I\end{bmatrix}\right|\left|\begin{bmatrix} x_p \\ x_c\end{bmatrix}\right|^2 =: -\sigma(|(x_p, x_c)|^2),\end{aligned}$$

which implies the flow condition (8.15).
Proof of (8.16) Simply observe that x_p remains constant across jumps, therefore

$$V(x^+, \mu x_p) - V(x, \mu x_p) = \lambda((\delta^+)^2 - \delta^2) = -\delta^2 \leq 0,$$

where we used the fact that δ is reset to zero at each jump. □

8.5 Simulation Example

Consider the following entries for the matrices in (8.3):

$$A_p = \begin{bmatrix} -2 & 0 \\ 0 & -0.9 \end{bmatrix}, \quad A_{pd} = \begin{bmatrix} -1 & 0 \\ -1 & -1 \end{bmatrix}, \quad B_p = \begin{bmatrix} 1 \\ 1 \end{bmatrix}, \quad C_p = \begin{bmatrix} 1 & 1 \end{bmatrix}, \tag{8.22}$$

and the selection for controller (8.4):

$$A_c = \begin{bmatrix} -1 & 0 \\ 1 & 0 \end{bmatrix}, \quad B_c = \begin{bmatrix} 1 \\ 0 \end{bmatrix}, \quad C_c = \begin{bmatrix} 0 & -1 \end{bmatrix}, \tag{8.23}$$

which involves an integral action possibly to deal with constant references or disturbances. When interconnecting plant and controller, we obtain a continuous-time dynamics as in the upper equation of (8.9a) with

$$\begin{gathered} A_p = \begin{bmatrix} -2 & 0 \\ 0 & -0.9 \end{bmatrix}, \; A_{pd} = \begin{bmatrix} -1 & 0 \\ -1 & -1 \end{bmatrix}, \; B_pC_c = \begin{bmatrix} 0 & -1 \\ 0 & -1 \end{bmatrix}, \\ A_c = \begin{bmatrix} -1 & 0 \\ 1 & 0 \end{bmatrix}, \; B_cC_p = \begin{bmatrix} 1 & 1 \\ 0 & 0 \end{bmatrix}. \end{gathered} \tag{8.24}$$

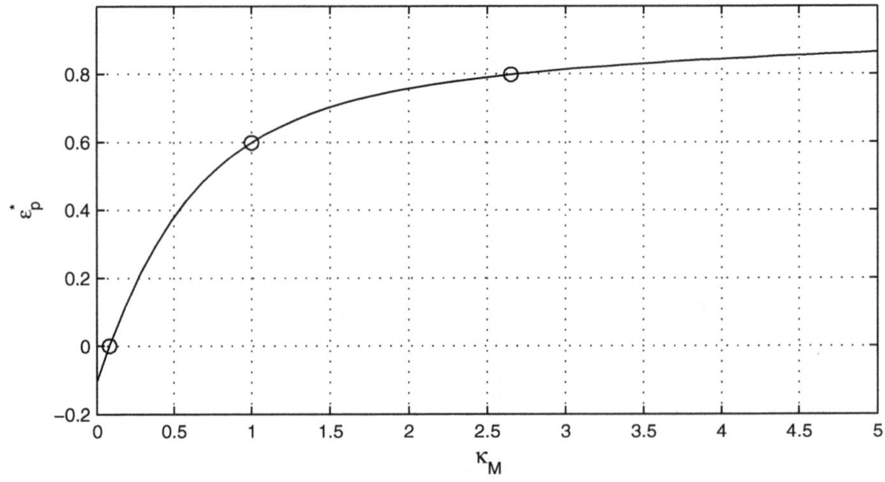

Fig. 8.1 The optimal value ε_p^* in (8.8) as a function of κ_M

Table 8.1 The three cases addressed in the simulation example

	κ_m	ε_p^*	K_p
Case 1	0.081	3.6×10^{-4}	$\begin{bmatrix} 0.0148 & -0.0001 \\ -0.0591 & -0.0542 \end{bmatrix}$
Case 2	1	0.5976	$\begin{bmatrix} 0.4501 & -0.1268 \\ -0.6102 & -0.7634 \end{bmatrix}$
Case 3	2.65	0.7977	$\begin{bmatrix} 1.3365 & -1.2358 \\ -1.3268 & -2.2350 \end{bmatrix}$

Such a continuous-time dynamics is exponentially stable if $\theta = 0$ (case with no delay) but for larger delays, beyond the critical value $\theta_M = 1.6$, the (linear) continuous-time closed loop becomes unstable.

Plant (8.3), (8.22) satisfies Assumption 1, therefore we may follow our hybrid construction to obtain GAS of the origin for any value of the delay, while preserving the continuous-time dynamics induced by (8.4), (8.24). To this aim, we follow the design paradigm in Remark 1 and compute a trade-off curve between the bound $\kappa_M > 0$ and the decrease rate ε_p^*.

Figure 8.1 shows the optimal values of ε_p^* as a function of the bound κ_M on $|K_p|$. Note that if κ_M is too small, then ε_p^* is negative and the design cannot be performed. Clearly, the curve is nondecreasing as increasing κ_M one enlarges the feasible set.

Table 8.1 shows the optimal values corresponding to the three circles reported in Fig. 8.1. The first value is just after the stability limit and the two other ones correspond to different trade-offs between ε_p^* and κ_M. For these three cases we run a time simulation selecting $\theta = 2$.

The simulation results for cases 1, 2, and 3, respectively, are shown in Figs. 8.2 and 8.3 using solid, dashed, and dotted curves, respectively. In the two figures, we

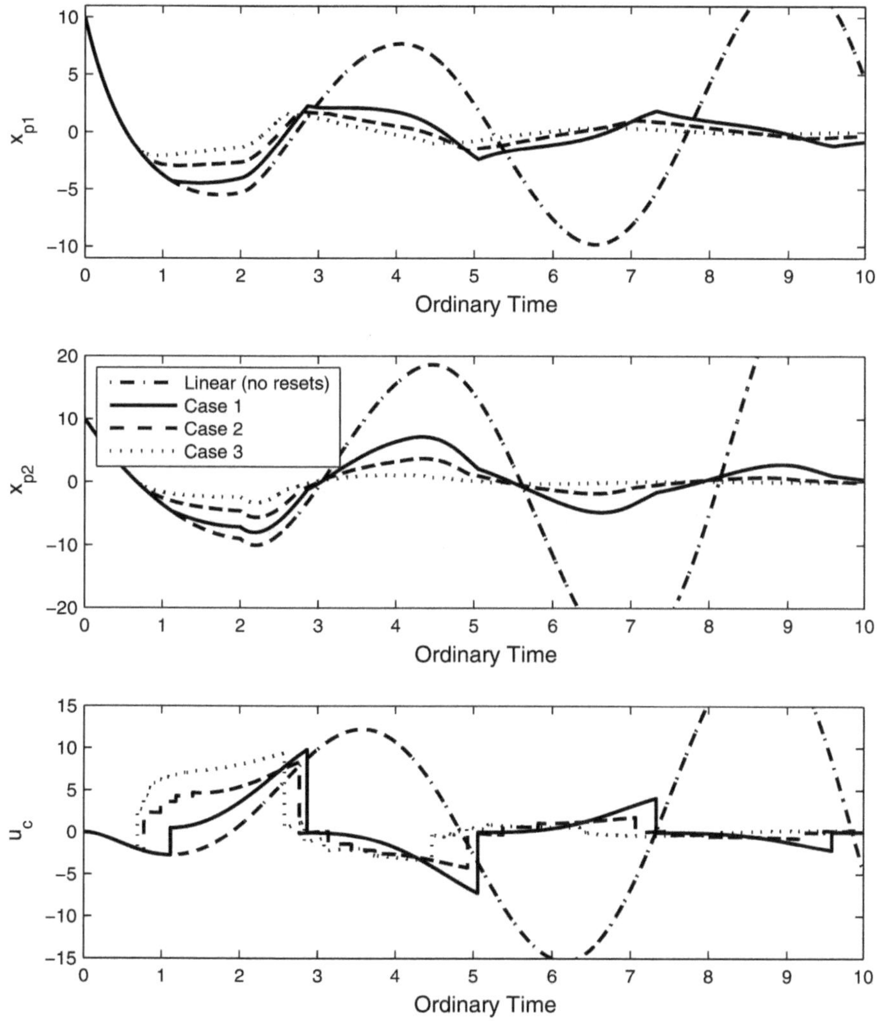

Fig. 8.2 States and input of the plant for the simulation test

also show the linear response (no resets), which is diverging (dash-dotted curves) because $2 > \theta_M \approx 1.6$. For the simulations, we selected $\rho = 0.01\ s$ that turns out to be well below the time elapsed between any pair of consecutive resets. Therefore, with reference to Remark 3, the dwell-time logic does not prematurely terminate our solutions.

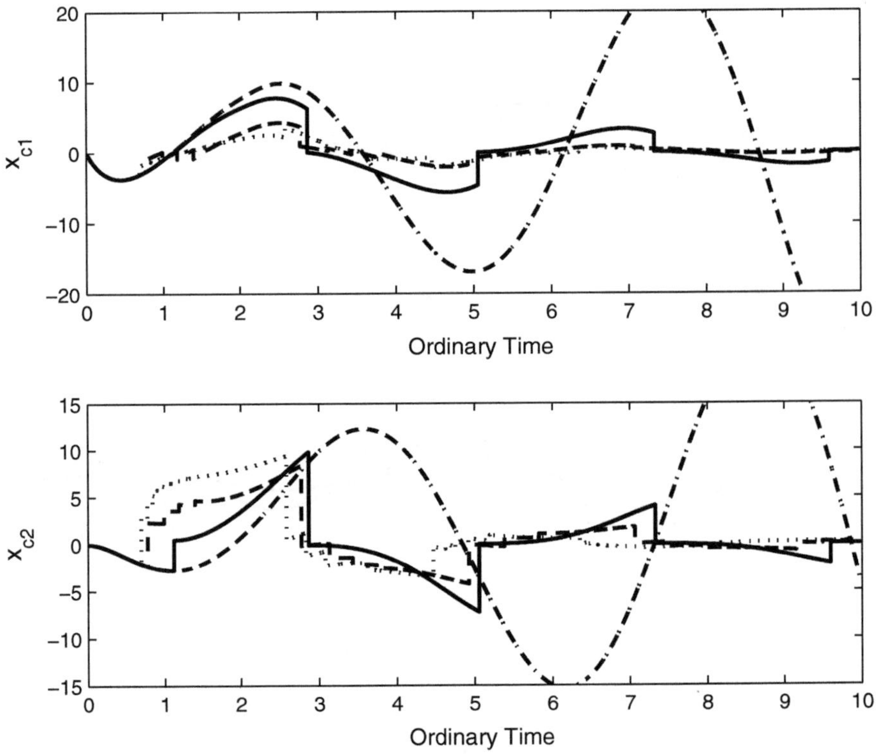

Fig. 8.3 States of the controller for the simulation test

The simulation results confirm the faster convergence rate (larger ε_p^*) envisioned for larger values of κ_M. Quite interestingly, this faster convergence is obtained by resetting earlier, rather than using a larger control input. Indeed, for larger values of κ_M, we observe a reduced amplitude of the control input u_c (see the bottom plot in Fig. 8.2).

Acknowledgments This work was supported by ANR project LimICoS contract number 12 BS03 005 01, by HYCON2 Network of Excellence "Highly- Complex and Networked Control Systems", grant agreement 257462 and by MICINN under the project DPI 2007-66455-C02-01.

References

1. Baños, A., Barreiro, A.: Delay-independent stability of reset systems. IEEE Trans. Autom. Control **54**(2), 341–346 (2009)
2. Baños, A., Barreiro, A.: Reset Control Systems. AIC Series, Springer, London (2012)
3. Baños, A., Pérez Rubio, F., Tarbouriech, S., Zaccarian, L.: Delay-independent stability via reset loops. LAAS-CNRS technical report (2013)

4. Barreiro, A., Baños, A.: Delay-dependent stability of reset systems. Automatica **46**(1), 216–221 (2010)
5. Clegg, J.C.: A nonlinear integrator for servomechnisms. Trans. A.I.E.E.m Part II **77**, 41–42 (1958)
6. Fichera, F., Prieur, C., Tarbouriech, S., Zaccarian, L.: Using Luenberger observers and dwell-time logic for feedback hybrid loops in continuous-time control systems. Int. J. Robust Nonlinear Control **23**, 1065–1086 (2013)
7. Goebel, R., Sanfelice, R.G., Teel, A.R.: Hybrid dynamical systems. IEEE Control Syst. Mag. **29**(2), 28–93 (2009)
8. Goebel, R., Sanfelice, R.G., Teel, A.R.: Hybrid Dynamical Systems: Modeling, Stability, and Robustness. Princeton University Press, Princeton (2012)
9. Guo, Y., Xie, L.: Quadratic stability of reset control systems with delays. In: World Congress on Intelligent Control and Automation, pp. 2268–2273 (2012)
10. Haddad, W.M., Chellaboina, V., Nersesov, S.G.: Impulsive and Hybrid Dynamical Systems: Stability, Dissipativity, and Control. Princeton University Press, Princeton (2006)
11. Hale, J.K., Lunel, S.M.V.: Introduction to functional differential equations. Applied Mathematical Sciences, vol. 99. Springer, New York (1993)
12. Horowitz, I.M., Rosenbaum, P.: Nonlinear design for cost of feedback reduction in systems with large parameter uncertainty. Int. J. Control **24**(6), 977–1001 (1975)
13. Liu, J., Teel, A.R.: Generalized solutions to hybrid systems with delays. In: Conference on Decision and Control, pp. 6169–6174. Maui, HI, USA (2012)
14. Nešić, D., Teel, A.R., Zaccarian, L.: Stability and performance of SISO control systems with first order reset elements. IEEE Trans. Autom. Control **56**(11), 2567–2582 (2011)
15. Nešić, D., Zaccarian, L., Teel, A.R.: Stability properties of reset systems. Automatica **44**, 2019–2026 (2008)
16. Prieur, C., Tarbouriech, S., Zaccarian, L.: Lyapunov-based hybrid loops for stability and performance of continuous-time control systems. Automatica **49**(2), 577–584 (2013)
17. Sipahi, R., Niculescu, S.-I., Abadallah, C.T., Michiels, W., Gu, K.: Stability and stabilization of systems with time delay. IEEE Control Syst. Mag. **31**(1), 38–65 (2011)
18. Sipahi, R., Vyhlídal, T., Niculescu, S.-I., Pierdomenico, P. (eds.): Time Delay Systems: Method, Applications and New Trends. Springer, New York (2012)
19. Zaccarian, L., Nešić, D., Teel, A.R.: Analytical and numerical Lyapunov functions for SISO linear control systems with first order reset elements. Int. J. Robust Nonlinear Control **21**(10), 1134–1158 (2011)

Chapter 9
Low Complexity Invariant Sets for Time-Delay Systems: A Set Factorization Approach

Sorin Olaru, Nikola Stanković, Georges Bitsoris and Silviu-Iulian Niculescu

Abstract This chapter deals with the study of invariant sets for discrete time linear systems affected by delay. It establishes a new perspective on their structural properties via set factorization. This novel perspective describes, in a unified framework, different existing notions of invariant sets. Additionally, it is shown that the (possible non-minimal) state space representation is a key element in the description of low complexity invariant sets.

9.1 Introduction

The present work focuses on systems described by discrete-time delay difference equations (dDDEs). The dimension of the state space and the delay values represent important parameters in the model, their values being the limiting factors for the application of classical linear time invariant (LTI) analysis and design methods.

S. Olaru (✉)
E3S (SUPELEC Systems Sciences) and EPI INRIA DISCO, 3 rue Joliot Curie, 91192 Gif-sur-Yvette, France
e-mail: sorin.olaru@supelec.fr

N. Stanković
E3S (SUPELEC Systems Sciences), 3 rue Joliot Curie, 91192 Gif-sur-Yvette, France
e-mail: nikola.stankovic@supelec.fr

G. Bitsoris
Control Systems Laboratory, Department of Electrical and Computer Engineering, University of Patras, 26500 Patras, Greece
e-mail: bitsoris@ece.upatras.gr

S.-I. Niculescu
Laboratory of Signals and Systems (L2S, UMR CNRS 8506), CNRS-Supelec, Gif-sur-Yvette, France
e-mail: silviu.niculescu@lss.supelec.fr

A. Seuret et al. (eds.), *Low-Complexity Controllers for Time-Delay Systems*, Advances in Delays and Dynamics 2, DOI: 10.1007/978-3-319-05576-3_9,

The analysis of time-delay systems is mainly based on he construction of Lyapunov like functions (functionals). In the case of linear discrete-time systems, the study is simplified by the fact that the difficulties due to the infinite dimensionality of the state space description are avoided. Nevertheless the computational complexity makes the analogy with the LTI systems impractical. This very simple remark points to the two available formalisms: Lyapunov-Krasovskii and Lyapunov-Razumikhin.

An analogy can be established for invariant set constructions for discrete-time Delay Difference Equations (dDDEs) where the two popular constructions are

- the invariant set in an extended state space on one side (extended state space whose dimension is in direct relation with the delay value)
- the invariant set in the original state space (which is independent of the delay value but often conservative as long as the existence conditions are restrictive).

In the present chapter the former will be called *extended invariance* while the later will be called *$\mathscr{D}$-invariance*. If the first concept is rather well understood and popular in the literature, the former has been formulated and (re)-investigated only recently.

The objective of the present work is threefold: (i) to show that $\mathscr{D}$-invariance can be seen as set-factorization of an invariant set in extended state space; (ii) to prove that these two constructions represent particular realizations of a broader family of invariant structures; (iii) to establish a connection between these families of invariant sets via set-factorization and similarity state transformations.

9.2 Basic Set Invariance Notions in Control

Given a mapping $f : \mathbb{R}^n \rightarrow \mathbb{R}^m$ and a set $\mathscr{S} \subseteq \mathbb{R}^n$, we define:

$$f(\mathscr{S}) = \{y \in \mathbb{R}^m | \exists x \in \mathscr{S}, \text{such that } y = f(x)\}.$$

Several definitions and properties on *positive invariance* are recalled next. Let us consider the discrete-time dynamical system:

$$x_{k+1} = f(x_k), \tag{9.1}$$

where $x_k \in \mathbb{R}^n$ is the state vector at time k and $f : \mathbb{R}^n \rightarrow \mathbb{R}^n$.

Definition 1 A set $\Omega \subset \mathbb{R}^n$ is positively invariant with respect to the dynamical system (9.1) if for any $x_0 \in \Omega$ it follows $x_k \in \Omega, \forall k \in \mathbb{Z}_+$. Alternatively, the set $\Omega \subset \mathbb{R}^n$ is positively invariant with respect to (9.1) if $f(\Omega) \subset \Omega$.

The positive invariance is a limit case of λ-contractiveness as indicated by the next definition and can be further elaborated for more general classes or dynamics or further related to stability notions (see [1]).

Definition 2 Given a scalar $0 \leq \lambda \leq 1$, a set $\Omega \subset \mathbb{R}^n$ containing the origin is called λ-contractive with respect to the dynamical system (9.1) if for any $x_0 \in \Omega$ it follows

$x_k \in \lambda^k \Omega, \forall k \in \mathbb{Z}_+$. Alternatively, the set $\Omega \in \mathbb{R}^n$ is λ-contractive with respect to (9.1) if $f(\Omega) \subset \lambda\Omega$.

The notions of positive invariance have been used in different research studies for analysis and control design. An important aspect of the invariant sets is the complexity of their representation. We recall next several constructions intended to provide alternatives to the classical Definition 2.

Definition 3 (Periodic invariance) For a given $\lambda \in \mathbb{R}_{[0,1)}$ the set $\Omega \subset \mathbb{R}^n$ containing the origin is called *periodic λ-contractive* with respect to the system (9.1) if there exists a positive number $p \in \mathbb{Z}_+$ such that for any $x_k \in \Omega$ it holds that $x_{k+p} \in \lambda\Omega$. If $\lambda = 1$ the set is called *periodic invariant*.

This periodic invariance has been proved to have a series of advantages for dynamics affected by parametric uncertainties or for switching systems [2].

Definition 4 (Cyclic invariance) For a given $\lambda \in \mathbb{R}_{[0,1)}$ the ordered family of sets containing the origin $\mathscr{O} = \left\{\Omega_1, \ldots, \Omega_p\right\} \subseteq \{\mathbb{R}^n\}^p$ is called *cyclic λ-contractive* with respect to (9.1) if for any $x_k \in \Omega_i$, $i \in \mathbb{Z}_{[1,p]}$ it holds that $x_{k+p} \in \lambda\Omega_i$ and $x_{k+j} \in \Omega_{i+j-\left\lfloor \frac{i+j}{p} \right\rfloor p}$ for $j \in \mathbb{Z}_{[1,p-1]}$.

This notion can be extended to the cyclic invariance or invariance over a family of sets.

Definition 5 (Invariant family of sets) For a given $\lambda \in \mathbb{R}_{[0,1)}$, the family of sets containing the origin $\mathscr{O} = \{\Omega_1, \ldots, \Omega_r\} \subseteq \{\mathbb{R}^n\}^r$ with $r \in \mathbb{Z}_+$ is called *λ-contractive* with respect to (9.1) if for any $x_k \in \Omega_i$, $i \in \mathbb{Z}_{[1,p]}$ it holds that $x_{k+p} \in \lambda\Omega_i$ and $x_{k+l} \in \bigcup_{j=1}^{r} \lambda\Omega_j$ for $l \in \mathbb{Z}_{[1,p-1]}$.

It should be noticed that there is a closed relation between these two definitions. In the case of cyclic invariance the sequence of transitions is pre-imposed in an ordered family of sets while in the case of an invariant family of sets the ordering is relaxed. From a theoretical point of view, the notion of invariant families of sets for the dynamical systems (9.1) does not introduce a novelty with respect to the Definition 2. Indeed, the invariance of the family of sets $\mathscr{O} = \{\Omega_1, \ldots, \Omega_r\}$ implies the invariance of the union $\mathscr{U} = \bigcup_{j=1}^{r} \Omega_j$. There is, however, a practical advantage of the invariant family if additional convexity restrictions are imposed to the members of the family. If the sets $\Omega_i, \forall i \in \mathbb{Z}_{[1,p]}$ composing the family $\mathscr{O}$ are convex, then the set $\mathscr{U} = \bigcup_{j=1}^{r} \Omega_j$ is not convex even if it is invariant. The interest of the invariant and cyclic invariant families of sets is revealed for distributed [3] and time-delay systems [4, 5] as detailed in the Sect. 9.3.

The invariant sets represent key elements in the constrained control design based on the so-called *receding horizon principle* (or Model Predictive Control—MPC). Consider the discrete-time system:

$$x_{k+1} = g(x_k, u_k), \tag{9.2}$$

where $g(0, 0) = 0$, $u_k \in U$ and $x_k \in X$ with U and X a convex, compact subsets of $\mathbb{R}^m$ and $\mathbb{R}^n$ respectively containing the origin in their interiors. Consider also the optimization problem

$$\begin{aligned} &\min_{u_k,\ldots,u_{k+N-1}} \sum_{i=k}^{k+N-1} l(x_i,u_i)+T(x_{k+N}),\\ &x_{i+1}=g(x_i,u_i), \forall i\in\mathbb{Z}_{[k,k+N-1]}\\ &x_i\in X,\ u_i\in U,\ x_{k+N}\in\Omega \end{aligned} \tag{9.3}$$

with $l(x,u)>0, l(0,0)=0$ and $l(x,u)>c|(x,u)|$, for some $c\in\mathbb{R}_+$, representing the stage cost in the objective function and $T(x)>0$, for all $x\neq 0$ and $T(0)=0$ as the terminal cost. The set $\Omega\in\mathbb{R}^n$ contains the origin in the interior and represents the terminal constraint. By applying the first component of the optimal control sequence $u_{mpc}=u_k^*$ and restarting the optimization, a feedback control law is obtained. The next results underlines the importance of the positive invariance for the stability of the resulting closed-loop system (see [6] for further details).

Proposition 1 *If there exists a function $\mathscr{K}:\Omega\rightarrow U$ such that:*

1. $g(x,\mathscr{K}(x))\in\Omega$, $\forall x\in\Omega$ *(positive invariance of Ω)*;
2. $T(g(x,\mathscr{K}(x)))+l(x,\mathscr{K}(x))-T(x)\leq 0$ *(local Lyapunov function).*

Then the control law u_{mpc} obtained by solving the receding horizon optimization (9.3) *guarantees the asymptotic stability of the resulting closed-loop system* (9.2).

The complexity of the predictive control law is related to the complexity of the optimization problem (9.3) and its terminal set. This remark underlines the importance of the complexity of invariant sets in the overall complexity of MPC. The Sect. 9.3 will come back on all the notions introduced up to this point and detail the analogies and the specificities of the time-delay dynamical systems.

9.3 Time-Delay Systems and Associated Invariant Sets

9.3.1 Dynamics Described by Discrete-Time Delay-Difference Equation

Consider the following delay-difference equation as a modeling framework for time-delay systems:

$$x_k=\sum_{i=1}^{m} A_i x_{k-d_i}, \tag{9.4}$$

with $A_i\in\mathbb{R}^{n\times n}$, $i=1,\ldots,m$ and initial state sequence $x_j\in\mathbb{R}^n$ for $j\in\mathbb{Z}_{[-d_m,-1]}$. The positive integers d_i, with $d_i<d_{i+1}$, called *delays*, form a vector $\mathbf{d}=[d_1\ldots d_m]^T\in(\mathbb{Z}_+)^m$ in the delay-parameter space.

For any delay vector $\mathbf{d}$, an extended state space representation can be constructed. For instance, if $d_{i+1}=d_i+1$, $d_1=1$, then, setting $X_k=\left[x_k^T\ldots x_{k-d_m+1}^T\right]^T$, the Eq. (9.4) is rewritten as:

$$X_k = A(\mathbf{d})X_{k-1} = \begin{bmatrix} A_1 & \dots & A_{d_m-1} & A_{d_m} \\ I & \dots & 0 & 0 \\ \vdots & \ddots & \vdots & \vdots \\ 0 & \dots & I & 0 \end{bmatrix} X_{k-1}, \tag{9.5}$$

where $\left[\mathbf{d} = 1 \dots d_m - 1\, d_m^T\right]$. With an appropriate re-ordering of the indices, and by setting $A_i = 0$ where necessary, we can obtain an extended state-space representation of (9.4) for any delay realization. Therefore, without loss of generality, the system matrix for the extended state space representation of (9.4) will be denoted by $A(\mathbf{d})$ according to the notation (9.5).

9.3.2 Set Invariance for Time-Delay Systems

As the main objective of this chapter is the study of the positive invariance for the dDDE (9.4), we introduce here a number of formal definitions of the existing notions in the literature and the reader is referred to [4, 7–9] for further details.

The direct transposition of the Definition 1 to the particular case of delay-time systems can be done in the extended state space (9.5).

Definition 6 Given a scalar $0 \leq \lambda \leq 1$, a set $\mathbb{S} \subseteq (\mathbb{R}^n)^{d_m}$ containing the origin is called λ-contractive with respect to the system (9.5) if for any $X_0 \in \mathbb{S}$ it follows $X_k \in \lambda^k \mathbb{S}$, $\forall k \in \mathbb{Z}_+$. Alternatively, the set $\mathbb{S}$ is λ-contractive with respect to (9.5) if $A(\mathbf{d})\mathbb{S} \subset \lambda\mathbb{S}$. In the case $\lambda = 1$ the set is called positively invariant.

It is worth mentioning that the fixed points and the periodic orbits are natural (and trivial) invariant sets. As such, for the linear dDDE case, the invariant sets contain the origin, and often the construction of non-trivial invariant sets starts from the assumption of having the origin as an interior point.

The positive invariant sets in the extended space may be very complex due to the dependence of the dimension of this state space on the maximal delay value d_m. In order to deal with a set invariance definition independent of this parameter, a notion close to the delay-independent stability has to be introduced.

Definition 7 ($\mathscr{D}$-invariance) Given a scalar $0 \leq \lambda \leq 1$, a set $\Omega \subset \mathbb{R}^n$ containing the origin is called $\mathscr{D}$-contractive with respect to the time-delay system (9.4) if for any initial conditions $x_i \in \Omega$, $i \in \mathbb{Z}_{[-d_m,-1]}$, it follows that $x_k \in \lambda\Omega$, $\forall k \in \mathbb{Z}_+$. Alternatively, the set Ω is $\mathscr{D}$-contractive with respect to (9.4) if $\bigoplus_{i=1}^{m} A_i\Omega \subset \lambda\Omega$. If $\lambda = 1$ the set is $\mathscr{D}$-invariant.

By analogy with the invariance notions introduced in Sect. 9.2, the cyclic invariance can offer a certain degree of flexibility by proposing a family of invariant sets instead of a rigid object in $(\mathbb{R}^n)^{d_m}$ or $\mathbb{R}^n$ as introduced in Definitions 6 and 7.

Definition 8 (Cyclic $\mathscr{D}$-invariance) A family of (d_m tuples of) sets $\{\Omega_1, \ldots, \Omega_{d_m}\}$ is called cyclic $\mathscr{D}$-invariant with respect to (9.4) if:

$$\begin{array}{ll} A_1\Omega_1 \oplus A_2\Omega_2 \oplus \cdots \oplus A_{d_m}\Omega_{d_m} & \subseteq \Omega_{d_m}; \\ A_1\Omega_{d_m} \oplus A_2\Omega_1 \oplus \cdots \oplus A_{d_m}\Omega_{d_m-1} & \subseteq \Omega_{d_m-1}; \\ \vdots & \\ A_1\Omega_2 \oplus A_2\Omega_3 \oplus \cdots \oplus A_{d_m}\Omega_1 & \subseteq \Omega_1. \end{array} \tag{9.6}$$

We conclude this series of invariance definitions by the generalization of the cyclic invariance to the invariant families of sets [4].

Definition 9 (*Invariant family* of sets for dDDE) A family of (d_m tuples of) sets $\mathscr{F} \subset (\mathbb{R}^n)^{d_m}$ is an invariant family with respect to (9.4) if for any tuple $\{\Omega_1, \Omega_2, \ldots, \Omega_{d_m}\} \in \mathscr{F}$ there exist a set $\Omega_0 \subset \mathbb{R}^n$ such that $\{\Omega_0, \Omega_1 \ldots, \Omega_{d_m-1}\} \in \mathscr{F}$ and

$$A_1\Omega_1 \oplus A_2\Omega_2 \oplus \cdots \oplus A_{d_m}\Omega_{d_m} \subseteq \Omega_0. \tag{9.7}$$

9.3.3 Positive Invariance and the Control Law Design

The MPC design can be reformulated for the delay-time systems:

$$x_k = \sum_{i=1}^{m} A_i x_{k-d_i} + B_i u_{k-d_i}, \tag{9.8}$$

subject to control and the state constraints $u_k \in U$ and $x_k \in X$ with U a convex, compact subset of $\mathbb{R}^m$ and X a convex, closed subset of $\mathbb{R}^n$, both containing the origin in their interior.

An MPC design based on the "terminal set"—"terminal cost" mechanisms for guaranteeing the stability can be formulated as

$$\begin{array}{l} \min\limits_{u_k,\ldots,u_{k+N-1}} \sum\limits_{i=k}^{k+N-1} l(x_i, u_i) + T(x_{k+N}), \\ x_i = \sum_{j=1}^{m} A_j x[i-d_j] + B_j u[i-d_j], \ \forall i \in \mathbb{Z}_{[k+1,k+N-1]} \\ x_i \in X, \ u_i \in U, \ \left[x_{k+N}^T \ldots x_{k+N-d_m+1}^T\right]^T \in \mathbb{S} \end{array} \tag{9.9}$$

where $\mathbb{S}$ is an invariant set in the extended state space as introduced in Definition 7.

Alternatively, the terminal constraint can be ensured by a formulation based on the existence of a $\mathscr{D}$-invariant set - Ω, as introduced in Definition 7:

$$x[k+N] \in \Omega, \ldots, x[k+n-d_m+1] \in \Omega. \tag{9.10}$$

Both formulations fulfill the original objective: enforcing the containment of the trajectories in a terminal invariant structure in order to ensure the stability of the control law. However, they present different characteristics, the $\mathcal{D}$-invariance being preferable from the complexity point of view, while the extended invariance describes generally a larger domain in the extended state space.

If the existence and construction of invariant sets in the extended state space is well understood and known to be related to the stability of the extended dynamics (9.5), the necessary and sufficient conditions for the existence of $\mathcal{D}$-invariant sets are still subject to research. In the Sect. 9.4 we present several geometrical notions related to set factorization which allow establishing a novel connection between extended and $\mathcal{D}$-invariance and further propose new invariant set structures.

9.4 Set Factorization

Given a set of indices $I \subset \mathbb{Z}_{[1,m]}$, a partition of I is described as a set of l subsets I_k of I such that $I = \bigcup\limits_{k=1}^{l} I_k$ with $I_i \cap I_j = \emptyset$. Given a subset $I_i \subset \mathbb{Z}_{[1,m]}$ and a set $P \in \mathbb{R}^m$, $P_{\downarrow I_i}$ denotes the projection of the set P on the subset of $\mathbb{R}^n$ with indices of Cartesian coordinates in I_i. Before introducing the main concepts it is useful to recall the definition of the Cartesian product of two sets P_1 and P_2:

$$P_1 \times P_2 = \{(x, y) | x \in P_1 \text{and } y \in P_2\}. \tag{9.11}$$

The next definitions of set factorization are closely related with the studies on space reduction for geometric computation [10, 11].

Definition 10 Let $\Omega \in \mathbb{R}^m$ and $\bigcup\limits_{i=1}^{l} I_i$ be the partition of $\mathbb{Z}_{[1,m]}$

1. The set Ω *is factorized* according to the partition $\bigcup\limits_{i=1}^{l} I_i = \mathbb{Z}_{[1,m]}$ if:

$$\Omega = \Omega_{\downarrow I_1} \times \cdots \times \Omega_{\downarrow I_l} \tag{9.12}$$

2. A *set factorization* (9.12) is *balanced* if $card\{I_1\} = \cdots = card\{I_L\}$.
3. A factorization is said to be *ordered* if it is defined by an *ordered* partition $I = \bigcup\limits_{k=1}^{l} I_k$ satisfying

$$\max\{I_i\} < \min\{I_j\}, \ \forall i < j; \tag{9.13}$$

4. A factorization is *regular* if is characterized by the equivalence of the factors

$$\Omega_{\downarrow I_1} = \cdots = \Omega_{\downarrow I_l} = P, \tag{9.14}$$

and

$$\Omega = \underbrace{P \times P \times \cdots \times P}_{l \text{ times}}. \tag{9.15}$$

Several remarks can be made with respect to the previous definitions:

- The set factorization is a non-commutative set operation (a Cartesian product inherited property). The regular factorization is one of the special cases where such commutativity property holds inside the given partition.
- Any regular factorization is balanced.
- The regularity of a factorization do not imply the ordering of the partition.
- A balanced and ordered factorization is not necessarily regular.

The general definitions of factorization do not imply a specific geometry of the set Ω. Most of the properties are related to the Cartesian product operation. It is worth mentioning, however, that the geometry of the factors can be related to the geometry of the set Ω. For example, the convexity of the set Ω implies the convexity of the factors and the polyhedral structure of factors implies a polyhedral structure for the set Ω. It should be noticed that even if the projection of ellipsoidal set is ellipsoidal, one cannot expect to obtain factorizations of ellipsoidal set as long as the Cartesian product of ellipsoids is not an ellipsoid.

From these remarks it becomes clear that the polyhedral sets represent a particularly interesting class which can be used as a framework for the developments in relationship with set factorization. We point the reader to the references [10, 11] and recall here a property of the polyhedral factorization:

Proposition 2 *There exists an ordered factorization of a polyhedral set described in its (non-redundant) half-space representation*

$$\Omega = \left\{x \in \mathbb{R}^n | Fx \leq w\right\}, \tag{9.16}$$

if and only if there exists a column permutation making the matrix F block diagonal.

Before describing the relationship between set factorization and positive invariance, we define the *minimal/maximal* factorization for any polyhedron P in terms of the number of factors. In this sense, the *minimal* factorization is defined by the trivial partition $I_1 = \mathbb{Z}_{[1,n]}$.

9.5 The Factorization Aspect of Set-Invariance

Starting from the definitions of the invariance in the extended state space we can observe that the dimension of the respective state space allows a regular factorization type of relationship with $\mathscr{D}$-invariance as formally stated in the next theorem.

Theorem 1 *The system* (9.4) *admits a non-trivial*[1] *$\mathscr{D}$-invariant set if and only if there exists an invariant set for* (9.5) *which admits a regular ordered factorization.*

Proof The necessity can be proved by observing that the existence of a $\mathscr{D}$-invariant set P with respect to (9.4) implies that if $x_{-d_i} \in P$, $1 \leq i \leq d_m$ then $x_0 \in P$.

[1] A nontrivial invariant set is understood as being non-degenerate, compact, convex and containing the origin in its interior.

Consequently, for any extended vector $X_{-1} = \left[x_{-1}^T \dots x_{-d_m}^T\right]^T$ contained in $\Omega = P \times P \times \dots \times P = P^d$ it holds that $X_0 \in \Omega$ which implies the positive invariance of Ω with respect to (9.5).

The sufficiency is slightly more involved. The regular ordered factorization implies the existence of a set $P = \Omega_{\downarrow \mathbb{Z}_{[1,n]}} = \Omega_{\downarrow \mathbb{Z}_{[n+1,2n]}} = \dots = \Omega_{\downarrow \mathbb{Z}_{[n(d-1)+1,nd]}}$. In order to prove its $\mathscr{D}$-invariance we write the first line of the extended dynamics (9.5) as:

$$x_k = \left[A_1 \dots A_{d_m-1}\, A_{d_m}\right] X_{k-1}. \tag{9.17}$$

By expressing the vector X_{k-1} in the extended form $X_{k-1} = \left[x_{k-1}^T \dots x_{k-d_m}^T\right]^T$ and exploiting the factorization we get a series of relations starting from the invariance property:

$$X_{k-1} \in P \rightarrow x_k \in \Omega_{\downarrow \mathbb{Z}_{[1,n]}}, \tag{9.18}$$

is equivalent with:

$$\left[A_1 \dots A_{d_m-1}\, A_{d_m}\right] X_{k-1} \in \Omega_{\downarrow \mathbb{Z}_{[1,n]}},\ \forall X_{k-1} \in \Omega, \tag{9.19}$$

or set-wise:

$$\left[A_1 \dots A_{d_m-1}\, A_{d_m}\right] \Omega \subset \Omega_{\downarrow \mathbb{Z}_{[1,n]}}. \tag{9.20}$$

Replacing the extended set by its factorized formulation we get:

$$\left[A_1 \dots A_{d_m-1}\, A_{d_m}\right] \left(\Omega_{\downarrow \mathbb{Z}_{[1,n]}} \times \dots \times \Omega_{\downarrow \mathbb{Z}_{[(d_m-1)n+1,d_m n]}}\right) \subset \Omega_{\downarrow \mathbb{Z}_{[1,n]}}, \tag{9.21}$$

which is equivalent to a Minkowski sum:

$$\bigoplus_{i=1}^{d_m} A_i \Omega_{\downarrow \mathbb{Z}_{[(i-1)n+1,in]}} \subset \Omega_{\downarrow \mathbb{Z}_{[1,n]}} = P, \tag{9.22}$$

and leads to the $\mathscr{D}$-invariance conditions for the set P

$$\bigoplus_{i=1}^{d_m} A_i P \subset P, \tag{9.23}$$

thus completing the proof. □

Theorem 2 *The system* (9.4) *admits a family of cyclic $\mathscr{D}$-invariant set in $\mathbb{R}^n$ if and only if there exists a family of cyclic invariant sets $\left\{\Omega_1, \dots, \Omega_{d_m}\right\} \subseteq \{\mathbb{R}^n\}^{d_m}$ with respect to* (9.5) *such that*

- *each set Ω_i, $i \in \mathbb{R}^n$ admits a balanced ordered factorization with the same partition of indices;*

- *these factorizations share the same family of d_m factors, their order inside the Cartesian product* (9.11) *being obtained via a circular shift.*

Proof Analogous to Theorem 1. □

A similar result can be stated for the invariant family of sets with respect to (9.4). In this case the balanced ordered factorizations will be done on a larger family of factors, the circular shift condition being relaxed.

From the results established so far it becomes clear that:

- The *extended state space invariance* corresponds to a *minimal* factorization. This is obvious since in this particular case there exists a single factor in the factorization, the set itself.
- Under the constraints imposed by the dimension of the original delay difference equation (9.4), the $\mathcal{D}$-invariance represents the *maximal* regular (ordered) factorization. In order to establish an invariance related result, under the assumption the state space model (9.4) is minimal, the dimension of the factors in the factorization process is lower bounded by n. Takeing into account that the extended state space (9.5) has dimension $(\mathbb{R}^n)^{d_m}$, the number of factors cannot be greater than d_m.

The regular ordered polyhedral set factorization passes by a block organisation of the set of constraints in the half space description as detailed in Proposition 2. A natural way of transforming the structure of a given polyhedral set $\Omega \in (\mathbb{R}^n)^{d_m}$ in view of factorization is based on linear similarity state transformations. The next result summarizes the degrees of freedom in this respect and its proof is a direct consequence of Proposition 2.

Proposition 3 *Let* $\Omega = \left\{X \in (\mathbb{R}^n)^{d_m} | FX \leq w\right\}$ *be an invariant set in the extended state space with respect to the system* (9.5). *A regular ordered factorization with dimension-n factors exists if there exists a transformation matrix* $T \in \mathbb{R}^{(nd_m)\times(nd_m)}$ *such that:*

$$TFT^{-1} = \begin{bmatrix} F_1 & 0 & \dots & 0 \\ 0 & F_2 & \ddots & 0 \\ \vdots & \ddots & \ddots & \vdots \\ 0 & \dots & 0 & F_{d_m} \end{bmatrix}. \tag{9.24}$$

Corollary 1 *Let a delay-difference equation described by* (9.4). *There exists a* $\mathcal{D}$*-invariant set for this dynamical system in* $\mathbb{R}^n$ *if the following conditions are fulfiled*:

- *There exists a similarity transformation T such that*

$$\begin{bmatrix} B_1 & \dots & B_{d_m-1} & B_{d_m} \\ I & \dots & 0 & 0 \\ \vdots & \ddots & \vdots & \vdots \\ 0 & \dots & I & 0 \end{bmatrix} = T \begin{bmatrix} A_1 & \dots & A_{d_m-1} & A_{d_m} \\ I & \dots & 0 & 0 \\ \vdots & \ddots & \vdots & \vdots \\ 0 & \dots & I & 0 \end{bmatrix} T^{-1}; \tag{9.25}$$

- *There exists an invariant set with respect to the system*

$$\tilde{X}_k = \begin{bmatrix} B_1 & \dots & B_{d_m-1} & B_{d_m} \\ I & \dots & 0 & 0 \\ \vdots & \ddots & \vdots & \vdots \\ 0 & \dots & I & 0 \end{bmatrix} \tilde{X}_{k-1}, \tag{9.26}$$

which admits a regular ordered factorization.

Proof The first condition introduce the similarity transformation in the construction of $\mathscr{D}$-invariant sets all by preserving the dynamical model in the form of a delay difference equation in $\mathbb{R}^n$. Following the results of the Proposition 3, such a similarity transformation represents a parametrization of the conditions for the existence of a regular ordered factorization. The second condition of the Corollary recalls the necessary and sufficient conditions in Theorem 1. □

With linear algebra manipulations it can be shown that the constraint imposed on the similarity transformation (9.25) is very restrictive[2] and allows only simple change of coordinates on the original delay difference equation, without a major impact on the regular ordered factorization.

This result has an interesting consequence as it opens the way for factorizations which are in between the *minimal* (extended state space invariant set) and the *maximal* (the $\mathscr{D}$-invariant set). The idea is to find a similarity transformation

$$\begin{bmatrix} B_1 & \dots & B_{r-1} & B_r \\ I & \dots & 0 & 0 \\ \vdots & \ddots & \vdots & \vdots \\ 0 & \dots & I & 0 \end{bmatrix} = T^{-1} \begin{bmatrix} A_1 & \dots & A_{d_m-1} & A_{d_m} \\ I & \dots & 0 & 0 \\ \vdots & \ddots & \vdots & \vdots \\ 0 & \dots & I & 0 \end{bmatrix} T, \tag{9.27}$$

which corresponds to a non-minimal state space equation (9.4) as long as $r < d_m$. A simple numerical example will illustrate in the Sect. 9.6 the fact that there might exist set-factorizations leading to invariant structures with an intermediate complexity in between $\mathbb{R}^n$ and $(\mathbb{R}^n)^{d_m}$. Such constructions can provide flexibility in the predictive control design described in Sect. 9.3.3.

9.6 Example of Factorization via Non-Minimal State Representation

Consider the delay difference equation:

$$x_{k+1} = x_k - 0.5x_{k-2}, \tag{9.28}$$

which does not possess a $\mathscr{D}$-invariant set. Its extended state realization:

---

[2] It practically holds for block diagonal matrices T.

$$P = \left\{ \begin{bmatrix} 0.7071 & -0.7071 \\ 0 & -1.0000 \\ -1.0000 & 0 \\ -0.7071 & 0.7071 \\ -0.4472 & 0.8944 \\ 0.4472 & -0.8944 \\ 0.0000 & 1.0000 \\ 1.0000 & 0.0000 \end{bmatrix} x \leq \begin{bmatrix} 0.2571 \\ 0.1970 \\ 0.3030 \\ 0.2571 \\ 0.2439 \\ 0.2439 \\ 0.1970 \\ 0.3030 \end{bmatrix} \right\}$$

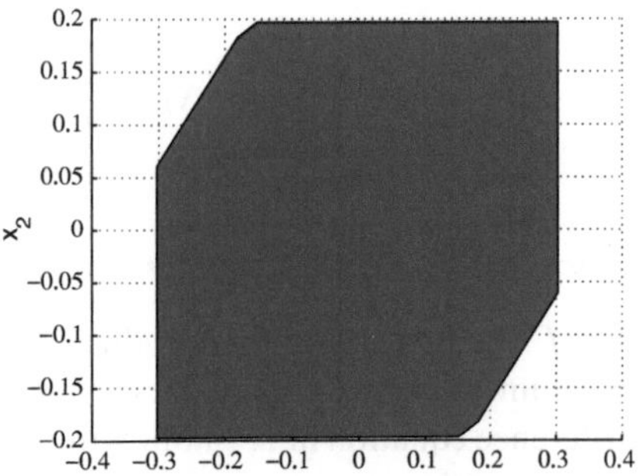

Fig. 9.1 $\mathcal{D}$-invariant set in a non-minimal state space

$$\begin{bmatrix} x_{k+1} \\ x_k \\ x_{k-1} \end{bmatrix} = \begin{bmatrix} 1 & 0 & -0.5 \\ 1 & 0 & 0 \\ 0 & 1 & 0 \end{bmatrix} \begin{bmatrix} x_k \\ x_{k-1} \\ x_{k-2} \end{bmatrix}, \tag{9.29}$$

has a strictly stable transition matrix and by consequence allows the construction of invariant sets in $\mathbb{R}^3$.

However these invariant sets with respect to (9.29) are not factorisable in $\mathbb{R}$ and following Theorem 1 cannot lead to $\mathcal{D}$-invariant sets with respect to (9.28). Interestingly, there exists a *nonminimal* state delay difference equation equivalent to (9.28):

$$\begin{bmatrix} x_{k+1} \\ x_{k+1} - x_k \end{bmatrix} = \begin{bmatrix} 0.5 & 0.5 \\ 0 & 0 \end{bmatrix} \begin{bmatrix} x_k \\ x_k - x_{k-1} \end{bmatrix} + \begin{bmatrix} 0 & 0.5 \\ -0.5 & 0.5 \end{bmatrix} \begin{bmatrix} x_{k-1} \\ x_{k-1} - x_{k-2} \end{bmatrix}, \tag{9.30}$$

for which a $\mathcal{D}$-invariant set $P \subset \mathbb{R}^2$ exists, an example being depicted in Fig. 9.1.

This will represent the factor of a invariant set in $\mathbb{R}^4$ with respect to the dynamics in a non-minimal state space:

$$\begin{bmatrix} x_{k+1} \\ x_{k+1} - x_k \\ x_k \\ x_k - x_{k-1} \end{bmatrix} = \begin{bmatrix} 0.5 & 0.5 & 0 & 0.5 \\ 0 & 0 & -0.5 & 0.5 \\ 1 & 0 & 0 & 0 \\ 0 & 1 & 0 & 0 \end{bmatrix} \begin{bmatrix} x_k \\ x_k - x_{k-1} \\ x_{k-1} \\ x_{k-1} - x_{k-2} \end{bmatrix}. \tag{9.31}$$

9.7 Conclusions

The chapter was dedicated to the positive invariance for discrete time-delay systems. It was shown that set-factorization is an key operation allowing the description of invariant sets in state-spaces of different dimensions. This flexibility opens the perspective for a better complexity management of the constraints describing the invariant sets. This is very important because low complexity invariant sets have implications in the complexity of the control design as for example in the MPC framework.

References

1. Blanchini. F., Miani, S.: Set-Theoretic Methods in Control. Springer, Birkhauser, Boston, (2008).
2. Lee, Y.I., Cannon, M., Kouvaritakis, B.: Extended invariance and its use in model predictive control. Automatica **41**(12), 2163–2169 (2005)
3. Rakovic, S.V., Kern, B., Findeisen, R.: Practical set invariance for decentralized discrete time systems. In: 49th IEEE Conference on Decision and Control, pp. 3283–3288 (2010).
4. Rakovic, S.V., Gielen, R.H., Lazar, M.: Construction of invariant families of sets for linear systems with delay. In: American Control Conference (ACC), pp. 6246–6251 (2012).
5. Lombardi, W., Olaru, S., Bitsoris, G., Niculescu, S.-I.: Cyclic invariance for discrete time-delay systems. Automatica **48**(10), 2730–2733 (2012)
6. Mayne, D.Q., Rawlings, J.B., Rao, C.V., Scokaert, P.O.: Constrained model predictive control: stability and optimality. Automatica **36**(6), 789–814 (2000)
7. Gielen, R., Lazar, M., Olaru, S.: Set-induced stability results for delay difference equations. In: Sipahi, R., Vyhlidal, T., Niculescu, S.-I., Pepe, P. (eds.) Time Delay Systems: Methods, Applications and New Trends, pp. 73–84. Springer (2012).
8. Lombardi, W., Olaru, S., Lazar, M., Niculescu, S.-I.: On positive invariance for delay difference equations. In: Proceedings of the IEEE American Control Conference, pp. 3674–3679 (2011).
9. Lombardi, W.: Constrained control for time-delay systems. Ph.D thesis, SUPELEC (2011).
10. Halbwachs, N., Merchat, D., Parent-Vigouroux, C.: Cartesian factoring of polyhedra in linear relation analysis. In: Cousot R. (ed.) Static Analysis. Lecture Notes in Computer Science, vol. 2694, pp. 355–365. Springer, Berlin Heidelberg (2003).
11. Halbwachs, N., Merchat, D., Gonnord, L.: Some ways to reduce the space dimension in polyhedra computations. Formal Methods Syst. Des. **29**(1), 79–95 (2006)

Chapter 10
Delays and Propagation: Control Liapunov Functionals and Computational Issues

Daniela Danciu and Vladimir Răsvan

Abstract There are considered some controlled objects with distributed parameters described by partial differential equations of hyperbolic type inducing wave propagation, connected at its turn with propagation delays. The boundary conditions are non-standard being described by ordinary differential or integro-differential equations. Basic theory—existence, uniqueness, well posedness-, stability and stabilization and numerical computations are considered for the benchmark problem of the marine vessel crane: its model is very much alike not only to other cranes but also to the flexible manipulator or the oilwell drillstring. In the lossless case basic theory is associated to the basic theory for some functional differential equations of neutral type. Stabilization is achieved by synthesizing low order controllers *via* c.l.f. (Control Liapunov Functional) induced by the *energy identity* for the partial differential equations. The numerics are considered within the framework of the *method of lines* implemented by applying the paradigm of the Cellular Neural Networks, an applied issue of Neuromathematics. After an illustrating simulation for the closed loop model some general conclusions and open problems are enumerated.

10.1 Motivating Applications and the Definition of a Class of Models

A. In a recent reference [9] we have described some apparently different models (i.e. from different fields of engineering) which share a common model. Without citing all references where these models occur (the interested reader may find their list within the reference list of the aforementioned chapter) we reproduce them below

- *The torque controlled flexible manipulator with distributed parameters*:

D. Danciu (✉) · V. Răsvan
Department of Automation, Electronics and Mechatronics, University of Craiova,
A.I. Cuza street 13, RO-200585 Craiova, Romania
e-mail: daniela@automation.ucv.ro

A. Seuret et al. (eds.), *Low-Complexity Controllers for Time-Delay Systems*,
Advances in Delays and Dynamics 2, DOI: 10.1007/978-3-319-05576-3_10,

$$
\begin{aligned}
y_{tt} - y_{xx} &= 0 \ , \ t > 0 \ , \ 0 < x < 1, \\
y_x(0,t) &= -\tau(t) \ , \ y_x(1,t) = 0, \\
\tau(t) &= -k_p \left[y(0,t) - \int_0^1 k(x) y_x(x,t) \mathrm{d}x \right] \\
& \quad -k_v \left[y_t(0,t) - \int_0^1 k(x) y_{tx}(x,t) \mathrm{d}x \right],
\end{aligned}
\tag{10.1}
$$

where the integration weight $k(x)$ admits the representation

$$
k(x) = -\int_x^1 g(\lambda) \mathrm{d}\lambda. \tag{10.2}
$$

This model may be viewed as a standard Neumann Boundary Value Problem (BVP). However, the control torque is designed using both a local (boundary) and a distributed feedback signal and the boundary conditions of the newly obtained system are of a different type.

The aforementioned model occurred by considering the flexible beam with negligible Young modulus $EI \approx 0$.

- *The controlled flexible arm of an ocean vessel riser.*
  Its model, deduced from the distributed parameter Hamilton principle, is endowed with disturbance signals, both distributed and lumped (i.e. at the boundaries) as well as with control signals (at the two boundaries as well):

$$
\begin{aligned}
& \rho y_{tt} - T y_{xx} + c y_t = f(x,t), \ t > 0, \ 0 < x < L, \\
& M \frac{\mathrm{d}^2}{\mathrm{d}t^2} y(0,t) + c_0 \frac{\mathrm{d}}{\mathrm{d}t} y(0,t) - T y_x(0,t) = u_0(t) + d_0(t), \\
& m \frac{\mathrm{d}^2}{\mathrm{d}t^2} y(L,t) + c_L \frac{\mathrm{d}}{\mathrm{d}t} y(L,t) + T y_x(L,t) = u_L(t) + d_L(t).
\end{aligned}
\tag{10.3}
$$

  Another difference with respect to (10.3) is the distributed damping $c > 0$; the significance of this non-negligible distributed damping will appear later.
- *The overhead crane with nonhomogeneous cable.*
  There is considered a slightly modified model (i.e. some simplifications are overruled), where the cable length is rated and a derivative boundary condition is introduced at $\sigma = 0$ by a non-negligible inertia as follows

$$
\begin{aligned}
& \frac{\rho L}{m} \cdot \frac{L}{g} y_{tt} - \left(\left(1 + \frac{\rho L}{m} \sigma \right) y_\sigma \right)_\sigma = 0, \ t > 0, \ 0 < \sigma < 1, \\
& \qquad \frac{L}{g} \frac{\mathrm{d}^2}{\mathrm{d}t^2} y(0,t) = y_\sigma(0,t) \ , \ y(1,t) = X_p, \\
& \qquad \frac{L}{g} \ddot{X}_p = \frac{m}{M} \left(1 + \frac{\rho L}{m} \right) y_\sigma(1,t) + \frac{L}{g} u(t).
\end{aligned}
\tag{10.4}
$$

- *The torsional stick slip oscillations in oilwell drillstrings.*
  There is consider a nonlinear model, the nonlinear function being introduced by the friction at the drilling head:

$$\begin{aligned} GJ\theta_{xx} - I\theta_{tt} - \beta\theta_t &= 0,\ t > 0,\ 0 < x < L, \\ GJ\theta_x(0,t) &= c_a(\frac{\mathrm{d}}{\mathrm{d}t}\theta(0,t) - \Omega(t)), \\ GJ\theta_x(L,t) + I_B\tfrac{\mathrm{d}^2}{\mathrm{d}t^2}\theta(L,t) &= -T(\tfrac{\mathrm{d}}{\mathrm{d}t}\theta(L,t)). \end{aligned} \tag{10.5}$$

B. In the following we shall discuss some specific features of the aforementioned models. All of them have what is called "distributed parameters" given by the element of the type "beam"—a mechanical elastic structure described by hyperbolic partial differential equations in "plane" (the time and one space dimension). In all these models the simplest beam—described by the vibrating string equation—has been used, but in different forms.

For instance, the beam of the flexible manipulator (10.1) is without any damping along it and uniform i.e. with space independent coefficients (and normalized). The drillstring and the vessel riser have distributed damping. The flexible cable of the overhead crane is nonuniform with respect to the space while without distributed damping. The nonuniformities will introduce however such effects that remind the distributed damping.

The boundary conditions of the above models also introduce some specific aspects. We shall discuss first the control theoretical aspects. The flexible manipulator has from this point of view also its specific properties: it is "closed loop" i.e. the possible control signals have been synthesized in order to ensure, *via* some feedback, stability and oscillation quenching. The other models are "open loop" i.e. the boundaries are subject to some input signals that can be either control signals or disturbances. It is the case of the overhead crane (10.4) as well as of the vessel riser (10.3). The drillstring has a control signal at one boundary only but it has another specific feature - the *nonlinear* torque at the drilling end.

On the other hand, the boundary conditions may be considered also from the purely mathematical point of view. Since they always contain some derivatives, their membership to the Neumann type could be considered but in fact the boundary dynamics may be of higher order or even of "Volterra operator" type, as mentioned in [1]. The boundaries of the flexible manipulator are a good example from this point of view.

C. For all these models the following mathematical problems may be formulated. There is first of all the so called *basic theory* which consists in proving existence of the solutions for given initial conditions, its uniqueness and well posedness in the sense of J. Hadamard. As pointed out by R. Courant [3], these properties ensure, within the process of modeling with Partial Differential Equations, the determinism and the possibility of making use of measurements and computations of limited precision without a "catastrophic behavior" of the model. We have to add here what is sometimes called "ubiquitous" character of the PDE - in this case the possibility of displaying smooth, discontinuous or even generalized solutions.

Such application problems as stability, stabilization, vibration quenching, disturbance rejection which are but too well known in the case of systems described by Ordinary or even Delay Differential Equations have their counterpart for PDE. We add to these problems of engineering analysis the synthesis problem - finding a feedback device (a controller) in order to ensure the aforementioned properties.

Finally, another application which is accompanying the modeling is the numerical computation. Here we have to mention another challenge—also known from the studies in ODE and DDE—*more sophisticated the numerical method is, smoother has to be the solution to ensure convergence*. Why is this a challenge in the case of the PDE, is more than obvious.

D. With the remark that controller synthesis may be, in principle, achieved at the formal level to obtain a closed loop model to consider for basic theory and for computations, we are in position to describe the structure of what is left of this contribution. After the introduction of the *energy identity*—well known in the theory of the PDE—a Liapunov functional allowing the synthesis of the feedback controller, this synthesis will be performed for the flexible riser model. The basic theory will be then tackled for the closed loop models considered in the chapter. Finally, the computational aspects will be considered within the *method of lines* which associates to the boundary value problem for PDE a system of ODE which is to be integrated. The advantage of the method of line is to associate an approximate system which is smooth enough for a good convergence. While high quality commercial software for the integration of the ODE is available, there will be proposed a computational structure inspired from the structure of the *Cellular Neural Networks* being thus more than adequate for the resulting large scale system of ODE. The chapter ends with conclusions and some perspective for future research problems.

10.2 Energy Identity. Liapunov Functional. Controller Synthesis

We shall deduce first the *energy identity*. Starting from (10.3), we multiply the first equation of the array by $y_t(x, t)$ and integrate from 0 to L along the solutions of (10.3) assumed to exist and to have integrable second order derivatives (in fact, everything in this section may be considered at the formal level up to the moment the controller synthesis is achieved and the closed loop system defined). After an integration by parts, the use of the boundary conditions and some rearrangement of the terms, the following identity is obtained

$$\frac{\mathrm{d}}{\mathrm{d}t}\frac{1}{2}\left[M\left(\frac{\mathrm{d}}{\mathrm{d}t}y(0,t)\right)^2 + m\left(\frac{\mathrm{d}}{\mathrm{d}t}y(L,t)\right)^2 + \int_0^L (\rho y_t^2(x,t) + Ty_x^2(x,t))\mathrm{d}x\right]$$

$$
\begin{aligned}
&+ c_0 \left(\frac{\mathrm{d}}{\mathrm{d}t} y(0,t)\right)^2 + c_L \left(\frac{\mathrm{d}}{\mathrm{d}t} y(L,t)\right)^2 + c \int_0^L \rho y_t^2(x,t)\mathrm{d}x \\
&- \frac{\mathrm{d}}{\mathrm{d}t} y(L,t)(u_L(t) + d_L(t)) - \frac{\mathrm{d}}{\mathrm{d}t} y(0,t)(u_0(t) + d_0(t)) \\
&- \int_0^L f(x,t) y_t(x,t)\mathrm{d}x \equiv 0
\end{aligned}
\tag{10.6}
$$

We take now into account the specific features of the control problem for the riser: the position of the payload has to be maintained closer to a given position P. For this reason the control Liapunov functional suggested by (10.6) is completed by a term accounting for the position error of the payload. This functional $\mathscr{V} : \mathbb{R} \times \mathbb{R} \times \mathbb{R} \times \mathscr{L}^2(0,L) \times \mathscr{L}^2(0,L) \mapsto \mathbb{R}_+$ reads

$$
\begin{aligned}
\mathscr{V}(X, Y, Z; \Phi(\cdot), \Psi(\cdot)) = \frac{1}{2}\Big[& k(X-P)^2 + MY^2 + mZ^2 \\
& + \int_0^L (\rho \Phi^2(x) + T\Psi^2(x))\mathrm{d}x \Big]
\end{aligned}
\tag{10.7}
$$

with the restriction $Y = \Phi(0)$; $Z = \Phi(L)$; also $k > 0$ is a free parameter. Obviously this functional is positive definite.

In order to synthesize the stabilizing controller, it is worth remembering that the Liapunov theory is concerned with *stability with respect to the initial conditions i.e. to short period perturbations*. We have thus to take for a while $f(x,t) \equiv 0$, $d_0(t) \equiv 0$, $d_L(t) \equiv 0$. The Liapunov functional (10.7) is written along the solutions of (10.3) as follows

$$
\begin{aligned}
&\mathscr{V}\left(y(L,t), \frac{\mathrm{d}}{\mathrm{d}t} y(0,t), \frac{\mathrm{d}}{\mathrm{d}t} y(L,t); y_t(\cdot,t), y_x(\cdot,t)\right) \\
&= \frac{1}{2}\Bigg[k(y(L,t)-P)^2 + M\left(\frac{\mathrm{d}}{\mathrm{d}t} y(0,t)\right)^2 + m\left(\frac{\mathrm{d}}{\mathrm{d}t} y(L,t)\right)^2 \\
&\qquad + \int_0^L (\rho y_t^2(x,t) + T y_x^2(x,t))\mathrm{d}x \Bigg].
\end{aligned}
\tag{10.8}
$$

Differentiating it along the solutions of (10.3) and taking into account the energy identity (10.6) as well as the assumptions on the perturbations we obtain

$$
\begin{aligned}
&\frac{\mathrm{d}}{\mathrm{d}t}\mathscr{V}\left(y(L,t), \frac{\mathrm{d}}{\mathrm{d}t} y(0,t), \frac{\mathrm{d}}{\mathrm{d}t} y(L,t); y_t(\cdot,t), y_x(\cdot,t)\right) \\
&= -c_0 \left(\frac{\mathrm{d}}{\mathrm{d}t} y(0,t)\right)^2 - c_L \left(\frac{\mathrm{d}}{\mathrm{d}t} y(L,t)\right)^2 - c \int_0^L \rho y_t^2(x,t)\mathrm{d}x \\
&+ \frac{\mathrm{d}}{\mathrm{d}t} y(L,t)(u_L(t) + k(y(L,t)-P)) + \frac{\mathrm{d}}{\mathrm{d}t} y(0,t) u_0(t).
\end{aligned}
\tag{10.9}
$$

This equality suggests the following local and *low order and low complexity* controllers

$$u_0(t) = -\phi_0\left(\frac{\mathrm{d}}{\mathrm{d}t}y(0,t)\right)\ ,\ u_L(t) = -k(y(L,t)-P) - \phi_L\left(\frac{\mathrm{d}}{\mathrm{d}t}y(L,t)\right), \quad (10.10)$$

where $\phi_i(\cdot)$, $i = 0, L$ are sector restricted nonlinear (in particular linear) functions, thus satisfying

$$0 < \frac{\phi_i(\sigma)}{\sigma} < \bar{\phi}_i\ ,\ \phi_i(0) = 0\ ,\ i = 0, L. \quad (10.11)$$

10.3 The Closed-Loop System and its Qualitative Properties

Substituting the controllers just synthesized in the equations (10.3), it is obtained the following boundary value problem with (possibly) nonlinear boundary conditions

$$\begin{aligned}
&\rho y_{tt} - Ty_{xx} + cy_t = f(x,t),\ t > 0,\ 0 < x < L,\\
&M\frac{\mathrm{d}^2}{\mathrm{d}t^2}y(0,t) + c_0\frac{\mathrm{d}}{\mathrm{d}t}y(0,t) + \phi_0\left(\frac{\mathrm{d}}{\mathrm{d}t}y(0,t)\right) - Ty_x(0,t) = d_0(t)\\
&m\frac{\mathrm{d}^2}{\mathrm{d}t^2}y(L,t) + c_L\frac{\mathrm{d}}{\mathrm{d}t}y(L,t) + k(y(L,t)-P) + \phi_L\left(\frac{\mathrm{d}}{\mathrm{d}t}y(L,t)\right) + Ty_x(L,t) = d_L(t)
\end{aligned} \quad (10.12)$$

with the associated positive definite Liapunov functional (10.7). As already seen, its derivative along the solutions of (10.12) is negative semi-definite.

Following [10] we can construct an abstract dynamical system associated to the unperturbed system (10.12), i.e. with $f(x,t) \equiv 0$, $d_0(t) \equiv 0$, $d_L(t) \equiv 0$, having the metric defined namely by the Liapunov functional (10.7). The invariance principle of Barbashin Krasovskii La Salle (BKL) can be applied to obtain global asymptotic stability of the steady state of the unperturbed system (10.12). This global asymptotic stability may be even exponential provided a construction of the Persidskii type [6] is valid in this case. This construction is realized using another Liapunov functional completely different from (10.7) and which is merely a proof instrument. At its turn, exponential stability may be used for the analysis of the behavior under perturbations—dissipativeness in the sense of N. Levinson (ultimate boundedness).

We shall not elaborate more on these topics which illustrate the possibilities of the actual theory of stability to provide various results by using Liapunov functionals with rather weak properties. We shall mention instead that stability constructions rely on the basic theory (existence, uniqueness well posedness [3]). From this point of view we already mentioned the rather general results of [1] which associate to (10.12) some functional equations of Volterra type. In the simpler case when $c = 0$ these equations are Neutral Functional Differential Equations and have been considered by many authors in studying basic theory, stabilization and other control aspects. *The computational issues for these equations are also interesting: if approximation of the delays by ordinary differential equations is considered in this case, the method of lines and its convergence is re-discovered for systems like* (10.12) [7]. *Implementation of the method of lines is considered in the next sections.*

10.4 Numerical Approach and Simulation

The problem taken into consideration in this section is that of (10.12), i.e. we have to solve numerically a mixed initial boundary value problem for a hyperbolic PDE. Our approach will combine the technique of the Method of Lines (MoL) with the peculiarities of a Cellular Neural Networks' structure which ensure the optimization of the storage and of the computational effort. It can be seen as a procedure which combines the classical computational mathematics with neurocomputing in order to solve an engineering application.

Neuromathematics [5], as a new branch of computational mathematics, aims to develop new methods and algorithms for solving both non-formalized (or weakly formalized) and formalized problems by using neural networks' paradigm. It basically relies on the fact that in the case of formalized problems (but not only) the weights of the interconnections not need a learning process based on experimental data in order to be achieved. In this case, the neural networks' structure is based mainly on the "natural parallelism" of the problem itself. A short list of formalized problems which can be solved using neurocomputing includes mathematical problems or tasks arising from natural sciences and engineering applications:

- high dimensional systems of linear or nonlinear algebraic equations and inequalities;
- optimization and systems identification problems;
- function approximation and extrapolation;
- (systems of) ordinary differential equations;
- partial differential equations.

As already said, in our approach we shall use first the Method of Lines for converting the specific initial boundary value problem for the hPDE (10.12) into a high-dimensional system of ODEs. Then, we shall use a Cellular Neural Network (CNN) structure for exploiting the local interconnections among the variables of the ODEs' system. This procedure preserves the qualitative properties of the initial problem and ensures the convergence of the approximation.

10.4.1 Applying the Method of Lines

Consider the controlled flexible arm of an ocean vessel riser (10.3), where the boundary control (10.10) was obtained via the control Liapunov functional (10.8) and replaced such that the closed loop dynamics is described by the equations (10.12). Taking $c = 0$, $d_0(t) = 0$, $d_L(t) = 0$ and $f(x,t) = 0$ and introducing $\sigma = x/L$—the rated (to the arm length L) arc length along the flexible non-stretching arm—the model becomes

$$
\begin{aligned}
& z_{tt}(\sigma,t) - \frac{T}{\rho L^2} z_{\sigma\sigma}(\sigma,t) = 0,\ t>0,\ 0<\sigma<1, \\
& M\ddot{z}(0,t) + c_0\dot{z}(0,t) + \phi_0(\dot{z}(0,t)) - \frac{T}{L} z_\sigma(0,t) = 0, \\
& m\ddot{z}(1,t) + c_L\dot{z}(1,t) + \phi_L(\dot{z}(1,t)) + \frac{T}{L} z_\sigma(1,t) + k(z(1,t)-P) = 0.
\end{aligned} \tag{10.13}
$$

Considering the new distributed variables $v(\sigma,t) = z_t(\sigma,t)$, $w(\sigma,t) = z_\sigma(\sigma,t)$ and denoting $z_0(t) = z(0,t)$, $z_1(t) = z(1,t)$ we obtain the system in the Friedrichs form

$$
\begin{aligned}
& v_t(\sigma,t) - \frac{T}{\rho L^2} w_\sigma(\sigma,t) = 0, \\
& w_t(\sigma,t) - v_\sigma(\sigma,t) = 0,
\end{aligned} \tag{10.14}
$$

with the boundary conditions (BCs)

$$
\begin{aligned}
& M\dot{v}(0,t) + c_0 v(0,t) + \phi_0(v(0,t)) - \frac{T}{L} w(0,t) = 0, \\
& m\dot{v}(1,t) + c_L v(1,t) + \phi_L(v(1,t)) + \frac{T}{L} w(1,t) + k(z_1(t)-P) = 0, \\
& \dot{z}_0(t) = v(0,t), \\
& \dot{z}_1(t) = v(1,t),
\end{aligned} \tag{10.15}
$$

For this system we introduce the Riemann invariants

$$
\begin{aligned}
r_1(\sigma,t) &= \frac{1}{2}\left(-v(\sigma,t) + aw(\sigma,t)\right), \\
r_2(\sigma,t) &= \frac{1}{2}\left(v(\sigma,t) + aw(\sigma,t)\right).
\end{aligned} \tag{10.16}
$$

The system (10.14)–(10.15) reads now as

$$
\begin{aligned}
& \frac{\partial r_1}{\partial t}(\sigma,t) + a\frac{\partial r_1}{\partial \sigma}(\sigma,t) = 0, \\
& \frac{\partial r_2}{\partial t}(\sigma,t) - a\frac{\partial r_2}{\partial \sigma}(\sigma,t) = 0,
\end{aligned} \tag{10.17}
$$

with the boundary conditions

$$
\begin{aligned}
& M(-\dot{r}_1(0,t) + \dot{r}_2(0,t)) - (c_0+b)r_1(0,t) + (c_0-b)r_2(0,t) \\
& \qquad + \phi_0(-r_1(0,t) + r_2(0,t)) = 0, \\
& m(-\dot{r}_1(1,t) + \dot{r}_2(1,t)) - (c_L-b)r_1(1,t) + (c_L+b)r_2(1,t) + k(z_1(t)-P) \\
& \qquad + \phi_L(-r_1(1,t) + r_2(1,t)) = 0,
\end{aligned}
$$

$$\begin{aligned} \dot{z}_0(t) &= -r_1(0,t) + r_2(0,t), \\ \dot{z}_1(t) &= -r_1(1,t) + r_2(1,t), \end{aligned} \tag{10.18}$$

where $a = \frac{1}{L}\sqrt{\frac{T}{\rho}}$ and $b = \sqrt{\rho T}$.

In order to apply the Method of Lines, we discretize the system (10.17)–(10.18) with respect to the spatial variable σ, by using the Courant-Isaacson-Rees rule with the Euler schemes. Thus, we shall obtain a good coupling with the boundary conditions and also, the convergence of the approximation—according to [7].

Consider the discretization step $h = 1/N$ of the interval [0, 1] and the notations $\sigma_i = ih$, $\xi_1^i(t) \approx r_1(\sigma_i, t)$ for $i = \overline{1, N}$ and $\xi_2^i(t) \approx r_2(\sigma_i, t)$ for $i = \overline{0, N-1}$. After some tedious but straightforward manipulations, we obtain the following systems of ordinary differential equations which embed the boundary conditions

$$\begin{aligned} &\dot{\xi}_1^0(t) = -A\xi_1^0 + B\xi_2^0 + s\xi_2^1 + \frac{1}{M}\phi_0(-\xi_1^0 + \xi_2^0), \\ &\dot{\xi}_1^i(t) = s\xi_1^{i-1} - s\xi_1^i, \quad i = \overline{1, N}, \\ &\dot{\xi}_2^i(t) = -s\xi_2^i + s\xi_2^{i+1}, \quad i = \overline{0, N-1}, \\ &\dot{\xi}_2^N(t) = -D\xi_2^N + E\xi_1^N + s\xi_1^{N-1} - \frac{k}{m}z_1 + \frac{k}{m}P - \frac{1}{m}\phi_L(-\xi_1^N + \xi_2^N), \end{aligned} \tag{10.19}$$

and

$$\begin{aligned} \dot{z}_0(t) &= -\xi_1^0 + \xi_2^0, \\ \dot{z}_1(t) &= -\xi_1^N + \xi_2^N, \end{aligned} \tag{10.20}$$

where the notations are: $s = aN, A = \frac{c_0+b}{M}, B = \frac{c_0-b}{M} - s, D = \frac{c_L+b}{m}, E = \frac{c_L-b}{m} - s.$

10.4.2 Implementation Using the CNNs' Paradigm

Cellular Neural Networks have some peculiarities of the structure which make them adequate for solving problems having some inherent parallelism and regularities as in the case of the tasks emerging when solving PDEs. Based on local interconnections only, one of the key idea of CNNs is that of representing the interactions among the cells by cloning templates—either translation-invariant or regularly varying templates [4]. To cite [2], "the cloning template is the elementary program of the CNN (an instruction or subroutine) and it can be specified by giving prescribed values, and/or characteristics to the template elements". There the authors referred to the case of programmability in VLSI implementation, but the structure of a CNN can be software implemented as well. The dynamics of a cell can be described by

$$\begin{aligned} \dot{x}_{ij} &= -x_{ij} + \sum_{kl \in N_r(ij)} A_{ij,kl}y_{kl} + \sum_{kl \in N_r(ij)} B_{ij,kl}u_{kl} + I_{ij}, \\ y_{ij} &= f(x_{ij}), \end{aligned} \tag{10.21}$$

where the notations are as follows: x_{ij} —the state variable of the cell ij, y—the output variable, u—the input (control) variable, $N_r(ij)$—the r—neighborhood of the cell ij, $A_{ij,kl}$—the feedback cloning template, $B_{ij,kl}$—the control cloning template, I_{ij}—the bias or an external input. The nonlinearity $f(\cdot)$ is the unit bipolar ramp function, but in some cases one can consider only the linear part, i.e. $f(x_{ij}) = x_{ij}$.

In order to reveal the so-called "cloning templates" or, more specific the feedback cloning templates—A_i, we rearrange the equations (10.19). Choosing as a new state vector

$$\zeta^T = [\xi_1^0 \; \xi_1^1 \; \ldots \; \xi_1^N \; \xi_2^N \; \xi_2^{N-1} \; \ldots \; \xi_2^1 \; \xi_2^0]. \tag{10.22}$$

and considering only the linear part of the nonlinearity $f(\cdot)$ we can embed the decay term in (10.21) within the feedback matrix A in order to obtain the inner cloning templates A_1 with $r = 1$ and A_2 with $r = 2$ and the general forms

$$\begin{aligned} A_1 &= [s \quad -s \quad 0], \qquad\quad i = \overline{2, N+1} \cup \overline{N+3, 2N+1}, \\ A_2 &= [s \quad -D \quad E \quad 0 \quad 0],\; i = N+2. \end{aligned} \tag{10.23}$$

The dynamics of these cells is described by the following equations

$$\begin{aligned} \dot{\zeta}_i &= \begin{pmatrix} s & -s & 0 \end{pmatrix} \begin{pmatrix} \zeta_{i-1} \\ \zeta_i \\ \zeta_{i+1} \end{pmatrix}, i = \overline{2, N+1} \cup \overline{N+3, 2N+1} \\ \dot{\zeta}_{N+2} &= \begin{pmatrix} s & -D & E & 0 & 0 \end{pmatrix} \begin{pmatrix} \zeta_N \\ \zeta_{N+1} \\ \zeta_{N+2} \\ \zeta_{N+3} \\ \zeta_{N+4} \end{pmatrix}. \end{aligned} \tag{10.24}$$

Let us emphasize that for the interconnection matrix A this arrangement leads to a lower two-diagonal matrix with some features of a circulant matrix (except for the rows $i = \{1, N+2\}$). These facts have important consequences on the computational and storage aspects.

The Eq. (10.20) can be viewed as another neural network which interacts with the first one, the CNN. The control inputs for the CNN are those which act on the boundary of the physical system ($x = 0, x = L$), i.e. on the cells ζ_1 and ζ_{N+2}.

10.4.3 Simulation Results

As usual, simulations are performed in order to verify the effectiveness of a method or of some theoretical results. We shall test our approach related to the properties of the approximated solution obtained for the mixed initial boundary value problem for a hPDE with nonlinear derivative BCs.

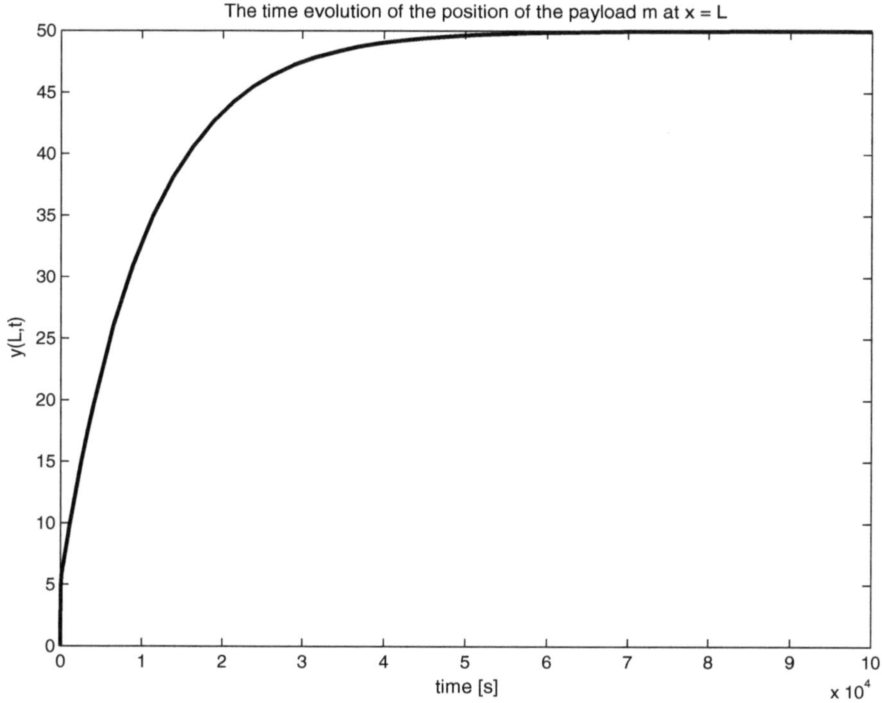

Fig. 10.1 The time evolution of the position of the payload m at $x = L$

The simulations were performed for the system of ODEs (10.19)–(10.20) which approximates the hPDE problem (10.13) for the controlled flexible arm of an ocean vessel riser. We have used the software implementation of the Cellular Neural Network with the state vector (10.22) and the feedback cloning templates (10.23) in order to make use of the sparsity induced by the PDE problem.

The values of the parameters are [8]: $g = 9.81$ m/s^2, $L = 1000$ m, $\rho = 108.02$ kg/m, $T = 8.11 \cdot 10^7$ N, $M = 9.6 \cdot 10^7$ kg, $m = 4 \cdot 10^5$ kg, $c_0 = 9 \cdot 10^7$ N $\cdot$ s/m, $c_L = 2 \cdot 10^5$ N $\cdot$ s/m, $P = 50.00$ m. The initial conditions are zero, i.e. the position and the velocity of the riser are zero at the initial time. The nonlinearities $\phi_0(\cdot)$ and $\phi_L(\cdot)$ are sigmoidal functions - the hyperbolic tangent function. In our case, the number of discretizing points $N = 10$ was sufficiently large for the convergence of the algorithm.

The simulations were performed by using the MATLAB ode15s solver in order to take advantage of the sparsity feature of the interconnection matrix of the CNN structure. The Fig. 10.1 presents the time evolution of the position of the payload m, i.e. the bottom boundary of the physical system, $x = L$. It can be seen that the target position $P = 50.00$ m is attained asymptotically. Also, Fig. 10.2 shows the space-time evolution of the velocity $v(\sigma, t) = z_t(\sigma, t) = y_t(x, t)$ of the flexible arm. Analyzing these plots we can conclude that the approximations obtained through

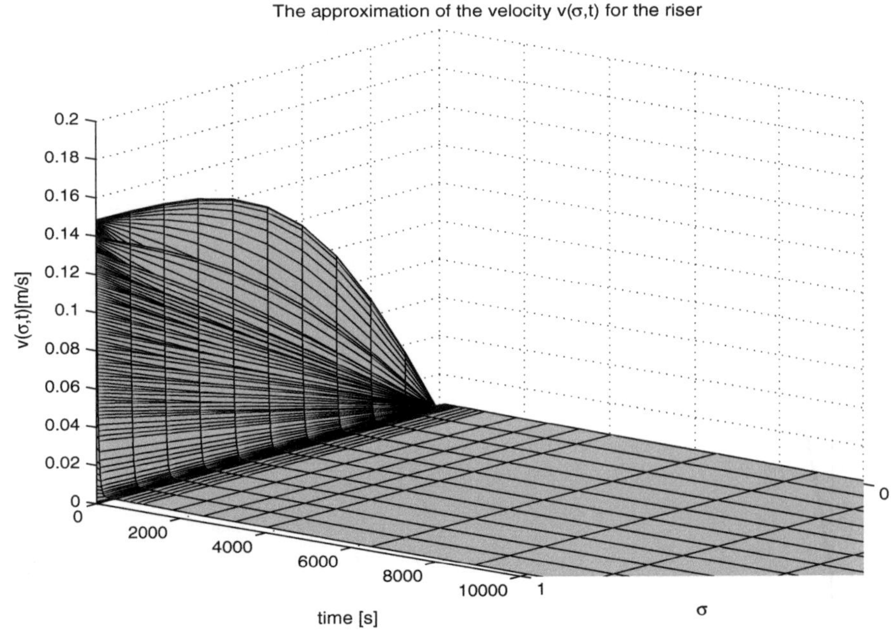

Fig. 10.2 3-D plot for the velocity $v(\sigma, t)$ of the flexible arm

our procedure fulfil the requirement of stability preserving for the initial system. Also, it can be seen that the procedure ensures the convergence of the approximation for $N = 10$ points used for the discretization of the domain of variation for the normalized space coordinate σ.

10.5 Some Conclusions and Open Problems

A. We have considered throughout this chapter some models of controlled objects with distributed parameters: cranes, oilwell drillstrings, flexible manipulators. Their models have in common the presence of a mechanical device of the type vibrating string described by the corresponding partial differential equation of hyperbolic type generating wave propagation and propagation delays. Either open loop (i.e. with control signals at the boundaries) or closed loop (i.e. with a synthesized boundary controller), the specificity of these models is given by dynamic boundary conditions i.e. described by differential, integral or integro-differential equations. For such "unusual" i.e. non-standard boundary value problems the basic theory should be the first step in model analysis and validation. In the case of lossless or distortionless propagation the entire basic theory (existence, uniqueness and well posedness) can be performed by associating some functional differential equations (in most cases of neutral type) for which the same basic theory is known. The association is in fact performed by integrating the Riemann invariants along the characteristics and its significance is a one-to-one correspondence between the solutions of the two mathematical objects. If the propagation is neither lossless nor distortionless, one has to rely on the very general framework of [1] without knowing too well how the Volterra operators mentioned there really look like. Therefore *construction of the specific basic theory for the aforementioned models is an urgent task*; like in the lossless/distortionless cases this theory may account for continuous and discontinuous classical solutions as well as for the generalized ones.

B. A second problem that has been tackled throughout the chapter has been the synthesis of low order boundary controllers using a special c.l.f. (Control Liapunov Functional). This functional is quadratic and generated by the *energy identity* from the theory of partial differential equations. The interesting (but not surprising fact) is that this synthesis may be performed at the formal level since any Liapunov like construction is valid along the solutions for which this functional is well defined. Once the controller being synthesized and substituted in the basic equations, the closed loop model is obtained and *it is this boundary value problem* that must be analyzed from the point of view of the basic theory.

On the other hand having at the disposal of a very "natural" energy-like Liapunov functional enables the analysis with a powerful tool of qualitative analysis: exponential stability using the Persidskii principle, ultimate boundedness results for systems subject to persistent perturbations are hints for other open research problems. The scientific progress is given here by getting all these properties *via* a Liapunov functional which has rather weak properties.

C. We would like to end this section by a discussion concerning the numerical studies. Among all methods developed for partial differential equations, the *Method of Lines* has as basic advantage the possibility of using all achievements (including software) of the field of Ordinary Differential Equations. For the hyperbolic case its convergence is not an easy task due to discontinuity propagation. Association of the functional differential equations allows replacement of this problem by the convergence of *delay approximation by ordinary differential equations* [7] whose convergence is valid even for discontinuous solutions; its counterpart is convergence of the method of lines for discontinuous classical and generalized solutions. Basic theory in the sense of [1] is for sure connected with the convergence of the method of lines in the corresponding case.

In this chapter we introduced another advantage of the method of lines in this case: discovery of some repetitive structures allowing to borrow the Cellular Neural Networks approach from the field of Neuromathematics. The present case may be viewed as a *toy application* but it is hoped to generate more efficient computational approaches in applications that are by far more complicated.

To summarize - an entire research programme to be followed.

Acknowledgments This work was partially supported by the grant number 10C/2014, awarded in the internal grant competition of the University of Craiova (Daniela Danciu) and by the project CNCS-Romania PN-II-ID-PCE-3-0198 (Vladimir Răsvan).

References

1. Abolinia, V.E., Myshkis, A.D.: Mixed problem for an almost linear hyperbolic system in the plane (in russian). Matem. Sbornik **50, 92**(4), 423–442 (1960)
2. Chua, L.O., Roska, T.: The cnn paradigm. IEEE Trans. Circuits Syst. I **40**(3), 147–156 (1993)
3. Courant, R.: Hyperbolic partial differential equations and applications. In: Beckenbach, E.F. (ed.) Modern Mathematics for the Engineer First Series, pp. 92–109. McGraw-Hill, New York (1956)
4. Danciu, D.: A cnn based approach for solving a hyperbolic pde arising from a system of conservation laws—the case of the overhead crane. In: Proceedings of 12th International Conference on Artificial Neural Networks: Advances in computational intelligence IWANN 2013, LNCS, pp. 365–374. Springer, Tenerife, Spain (2013)
5. Galushkin, A.I.: Neural Network Theory. Springer, Heidelberg (2010)
6. Halanay, A.: Differential Equations: Stability, Oscillations, Time Lags, Mathematics in Science and Engineering, vol. 23. Academic Press, New York and London (1966)
7. Halanay, A., Răsvan, V.: Approximations of delays by ordinary differential equations. In Conti, R. (ed.) Recent Advances in Differential Equations, pp. 155–197. Academic Press, New York London Toronto Sydney (1981)
8. He, W., Zhang, S., Ge, S.S.: Boundary control of a flexible riser with the application to marine installation. IEEE Trans. Ind. Electr. **60**(12), 5802–5810 (2013)
9. Răsvan, V.: Liapunov functionals for systems with time delay and propagation. In: proceedings of 2nd International Conference on Systems and Computer Science ICSCS2013, pp. 177–182. Villeneuve d'Ascq, France, 26–27 Aug 2013
10. Saperstone, S.H.: Semidynamical Systems in Infinite Dimensional Spaces, Applied Mathematical Sciences, vol. 37. Springer, New York Heidelberg Berlin (1981)

Chapter 11
Integral Inequality for Time-Varying Delay Systems and Its Application to Output-Feedback Control

Alexandre Seuret and Frédéric Gouaisbaut

Abstract This chapter considers the stability and stabilization of time-varying delay systems. We develop some new integral inequalities which are proved to encompass the celebrated Jensen's inequality. These technical tools allow to construct simple Lyapunov-Krasovskii functionals very efficient in practice. Notice that our procedure is coupled with the use of the reciprocal convexity result in order to reduce the conservatism induced by the LMI optimisation setup. The effectiveness of the proposed results is illustrated by an example extracted from the dynamics of machines chatter.

11.1 Introduction

The delay phenomenon arises in many practical situations like in networked control systems, biology, economy or mechanical engineering (see for instance [5, 19]) and references therein) and therefore, since several years, it has motivated a huge number of results devoted to the construction of stability and stabilization criteria for linear time delay systems.

For the case of time-varying delay, the results are often conservative and rely generally on two different methodologies. In the first framework, we aim at transforming the original system into a closed loop between a nominal system and a system depending on the delay. This last element is embedded into an uncertainty and the use of classical tools like Small Gain Theorem [4, 12], IQCs [8] or Quadratic Separation [2] allows to conclude on the stability. The proposed modeling is then employed to derive robust controller with respect to the delay (see for instance [14] and references therein). The another technique is the extension of Lyapunov theory to

A. Seuret (✉)
CNRS, LAAS, Université de Toulouse, 7 avenue du Colonel Roche, 31077 Toulouse, France
e-mail: aseuret@laas.fr

F. Gouaisbaut
Université de Toulouse, UPS, LAAS, 31400 Toulouse, France
e-mail: fgouaisb@laas.fr

A. Seuret et al. (eds.), *Low-Complexity Controllers for Time-Delay Systems*, Advances in Delays and Dynamics 2, DOI: 10.1007/978-3-319-05576-3_11,

the case of delay systems. In a certain sense, all the results follow the same methodology [5]. Firstly, the initial retarded system is transformed into a more suitable equation. This transformation may not be equivalent with regards to the stability: the original system may be stable but not the transformed system [5]. Then, the structure of a Lyapunov-Krasovskii functional is proposed, of which aim is to prove the stability of the transformed system and the original one as well. This choice of this structure *a priori* often induces an important conservatism and several attempts have been proposed to reduce it by choosing extended state based Lyapunov-Krasovskii functional [1, 9], or discretized Lyapunov functional [5]. The third step which brings an important source of conservatism comes from the inequalities to be used to derive a tractable numerical optimization problem (see [6, 7, 10, 15, 18, 20]). Generally, all the articles proceed in two stages: Firstly, all the cross terms of the derivative of V are bounded using a Jensen's like inequality. Then, in order to get an LMI with respect to the delay $h(t)$ and the delay derivative $\dot{h}(t)$, the result is transformed into a convex optimization problem often with the help of slack variables [15].

At this stage, we obtain an analysis LMI problem which can easily be turned into a synthesis problem by considering the stability of the closed loop. We aim therefore at finding the best controller gain which maximizes the stability region. Nevertheless, considering that the controller gain is unknown results often into a non linear optimization problem. In that case, a last step is often required to transform it into a linear problem either by fixing some decisions variables or by finding linearisation procedure [13] (re-introducing relaxation variables like in [15]).

In this chapter, unlike many articles in the literature, we focus on the these two last stages of the procedure, the development of less restrictive inequalities and the choice of slack variables. First of all, following the work of [3] or [10], we propose to construct more accurate inequalities than Jensen's or Wirtinger's ones. The resulting inequalities depend not only on the state $x(t)$ and the delayed state but also on the integral of the state over a delay interval. This new signal is then directly integrated into a suitable classical Lyapunov function, highlighting so the features of this new inequality. Then, we propose to extend the work of [15] in order to get tractable optimization scheme but with a fewer numbers of slack variables to be optimized.

The chapter is organized as follows. Section 11.2 introduces a new integral inequality which can include the Jensen's one as a special case. We also propose a new reciprocally convex combination inequality in order to get some tractable convex optimisation problem to deal with. Section 11.3 is devoted to the use of classical Lyapunov-Krasovskii functional combined with inequalities describes in Sect. 11.2. Section 11.4 illustrates our results by considering the regenerative chatter stabilization problem [19].

Notations: Throughout the chapter $\mathbb{R}^n$ denotes the n-dimensional Euclidean space with vector norm $|\cdot|$, $\mathbb{R}^{n\times m}$ is the set of all $n\times m$ real matrices. For any symmetric matrix $P\in\mathbb{R}^{n\times n}$, the notation $P\succ 0$ (or $P\prec 0$) means that P is positive (or negative) definite. The set $\mathbb{S}_+^n$ refers to the set of symmetric positive definite matrices. The symmetric matrix $\begin{bmatrix} A & B \\ * & C \end{bmatrix}$ stands for $\begin{bmatrix} A & B \\ B^T & C \end{bmatrix}$. For any matrices A, B in

$\mathbb{R}^{n\times n}$, the notation $diag(A, B)$ denotes the bloc diagonal matrix $\begin{bmatrix} A & 0 \\ * & B \end{bmatrix}$. Finally, a set of scalar parameter $\alpha_1, \ldots, \alpha_N, N$ belongs to the set $\mathscr{D}$, if each parameter is positive and $\sum_{i=1}^{N} \alpha_i = 1$.

11.2 Some Relevant Inequalities

In this section, we aim at presenting several integral inequalities which will be employed in order to derive stability conditions for linear systems with time-varying delays. More specifically, we formulate a new inequality which is proved to be less conservative compared to Jensen's inequality. Furthermore, based on a reciprocally convex combination lemma provided in [15], we propose some new relaxations for this lemma which allow to reduce the number of variables involved in the optimization setup.

11.2.1 Improved Integral Inequality

In the following subsection, we develop a new integral inequality which introduces less conservatism than the well-known Jensen's inequality.

Lemma 1 *For a given symmetric positive definite matrix $R \in \mathbb{S}_n^+$, any differentiable function ω in $[a, b] \rightarrow \mathbb{R}^n$, then the following inequality holds*:

$$\int_a^b \dot{\omega}(u)R\dot{\omega}(u)\mathrm{d}u \geq \tfrac{1}{b-a}(\omega(b)-\omega(a))^T R(\omega(b)-\omega(a)) + \tfrac{12}{b-a}\Omega^T R\Omega. \quad (11.1)$$

where $\Omega = \frac{\omega(b)+\omega(a)}{2} - \frac{1}{b-a}\int_a^b \omega(u)\mathrm{d}u.$

Proof For any sufficiently differentiable function ω in $[a, b] \rightarrow \mathbb{R}^n$, consider a signal z given, for all $u \in [a,\ b]$ by

$$z(u) = \dot{\omega}(u) - \frac{1}{b-a}(\omega(b) - \omega(a)) + 6\frac{(b+a-2u)}{(b-a)^2}\Omega,$$

where Ω has been defined in the statements of the lemma. The computation of $\int_a^b z^T(u)Rz(u)\mathrm{d}u$ leads to:

$$\begin{aligned}\int_a^b z^T(u)Rz(u)\mathrm{d}u = &\int_a^b \dot{\omega}^T(u)R\dot{\omega}(u)\mathrm{d}u\\
&+\frac{\int_a^b 1\mathrm{d}u}{(b-a)^2}(\omega(b)-\omega(a))^T R(\omega(b)-\omega(a))\\
&+36\frac{\int_a^b (b+a-2u)^2\mathrm{d}u}{(b-a)^4}\Omega^T R\Omega\\
&-\frac{2}{b-a}\int_a^b \dot{\omega}^T(u)\mathrm{d}uR(\omega(b)-\omega(a))\\
&+\frac{12}{(b-a)^2}\int_a^b (b+a-2u)\dot{\omega}^T(u)\mathrm{d}uR\Omega\\
&-\frac{12}{(b-a)^3}\int_a^b (b+a-2u)\mathrm{d}u(\omega(b)-\omega(a))^T R\Omega. \qquad (11.2)\end{aligned}$$

Simple calculus ensure that $\int_a^b (b+a-2u)^2\mathrm{d}u = \frac{(b-a)^3}{3}$ and $\int_a^b (b+a-2u)$ $\mathrm{d}u = 0$. Then an integration by parts ensures that

$$\int_a^b (b+a-2u)\dot{\omega}^T(u)\mathrm{d}u = -2(b-a)\Omega^T.$$

It thus follows that

$$\begin{aligned}\int_a^b z^T(u)Rz(u)\mathrm{d}u = &\int_a^b \dot{\omega}^T(u)R\dot{\omega}(u)\mathrm{d}u\\
&+\frac{1}{(b-a)}(\omega(b)-\omega(a))^T R(\omega(b)-\omega(a))\\
&-\frac{2}{(b-a)}(\omega(b)-\omega(a))^T R(\omega(b)-\omega(a))\\
&+\frac{12}{(b-a)}\Omega^T R\Omega - \frac{24}{(b-a)}\Omega^T R\Omega. \qquad (11.3)\end{aligned}$$

Since the matrix R is positive definite, the left-hand side of the previous equation is positive definite. This allows us to conclude the proof of Lemma 1. □

In order to compare with the celebrated Jensen's Lemma, let us recall this lemma, often at the root of several important results on the stability of time delay systems:

Lemma 2 (Jensen's Lemma) *For given symmetric positive definite matrices $R > 0$ and for any differentiable signal ω in $[a,\ b] \to \mathbb{R}^n$, the following inequality holds:*

$$\int_a^b \dot{\omega}(u) R \dot{\omega}(u) \mathrm{d}u \geq \tfrac{1}{b-a} (\omega(b) - \omega(a))^T R (\omega(b) - \omega(a)). \tag{11.4}$$

Proof A proof can be found in [5, 14]. However the previous lemma allows to give an alternative proof by considering $\Omega = 0$ in the definition of the function z. □

Remark 1 The difference between the two inequalities are the following: the first term in Lemma 1 refers exactly to the right hand side of the Jensen's inequality provided in Lemma 2. The second term of (11.1) is positive definite. Thus it is clear that this new inequality encompasses the Jensen's inequality. It is also worth noting that this improvement is allowed by using an extra signal $\int_a^b \omega(u)\mathrm{d}u$ and not only the signals $\omega(b)$ and $\omega(a)$. Therefore, it suggests that in order to be useful, this inequality should be combined with a Lyapunov functional where the signal $\int_a^b \omega(u)\mathrm{d}u$ appears explicitly.

Remark 2 A similar integral inequality based on Wirtinger's inequality has been recently proposed in [16, 17]. The difference with respect to this inequality relies in the coefficient 12 which replaces the previously obtained coefficient π^2 (which is lower than 12). This new approach results in a less conservative result.

11.2.2 Improved Reciprocally Convex Combination Inequality

The following is largely inspired from the reciprocally convex combination lemma provided in [15]. Recall firstly an useful lemma provided in this chapter:

Lemma 3 *[15] Let $f_1, f_2, \ldots, f_N$: $\mathbb{R}^m \to \mathbb{R}$ have positive values in an open subset $\mathcal{D}$ of $\mathbb{R}^m$. The reciprocally convex combination of f_i over $\mathcal{D}$ satisfies*

$$\min_{\{\alpha_i\}_i} \in \mathcal{D}\} \sum_i \frac{1}{\alpha_i} f_i(t) = \sum_i f_i(t) + \max_{g_{ij}(t)} \sum_i \sum_{j \neq i} g_{ij}(t), \tag{11.5}$$

subject to

$$\left\{ g_{ij} : \mathbb{R}^m \to \mathbb{R},\ g_{ij}(t) \triangleq g_{ji}(t),\ \begin{bmatrix} f_i(t) & g_{ij}(t) \\ g_{ij}(t) & f_j(t) \end{bmatrix} \succeq 0 \right\}. \tag{11.6}$$

An application of this lemma which was already proposed in the same article is stated as follows.

Lemma 4 *Let n, m be two integers, R a symmetric matrices in* $\mathbb{S}^n$ *and two matrices* W_1 *and* W_2 *in* $\mathbb{R}^{n\times nm}$. *Consider a scalar* α *in the interval* $(0,\ 1)$. *For all vector* ξ *in* $\mathbb{R}^{nm}$, *define the function* $\Theta(\alpha, R)$ *as:*

$$\Theta(\alpha, R) = \frac{1}{\alpha}\xi^T W_1^T R W_1 \xi + \frac{1}{1-\alpha}\xi^T W_2^T R W_2 \xi.$$

Then the improved reciprocally convex combination guarantees that if there exists a matrix X in $\mathbb{R}^{n\times n}$ *such that*

$$\begin{bmatrix} R & X \\ * & R \end{bmatrix} \succ 0,$$

then the following inequality holds

$$\min_{\alpha\in(0,\ 1)} \Theta(\alpha, R) \geq \begin{bmatrix} W_1\xi \\ W_2\xi \end{bmatrix}^T \begin{bmatrix} R & X \\ * & R \end{bmatrix} \begin{bmatrix} W_1\xi \\ W_2\xi \end{bmatrix}.$$

Proof The proof is omitted but can be found in [15]. □

In the following, an alternative lemma is proposed in order to reduce the number of the slack variables by eliminating the matrix X.

Lemma 5 *Let n, m be two integers, R a symmetric matrix in* $\mathbb{S}^n$ *and two matrices* W_1 *and* W_2 *in* $\mathbb{R}^{n\times nm}$. *Consider a scalar* α *in the interval* $(0,\ 1)$. *Then the improved reciprocally convex combination guarantees that*

$$\min_{\alpha\in(0,\ 1)} \Theta(\alpha, R) \geq \xi^T \left[W_1 - W_2\right]^T R \left[W_1 - W_2\right] \xi.$$

Proof As mentioned in the introduction of the section, the proof is based on the main result of [15] but adapted to our case in order to reduce the number of slack variables. Following [15], for all ξ in $\mathbb{R}^{nm}$, the function $\Theta(\alpha, R)$ can be rewritten as

$$\begin{aligned}\Theta(\alpha, R) &= \tfrac{\alpha+(1-\alpha)}{\alpha}\xi^T W_1^T R W_1 \xi + \tfrac{\alpha+(1-\alpha)}{1-\alpha}\xi^T W_2^T R W_2 \xi \\ &= \xi^T W_1^T R W_1 \xi + \xi^T W_2^T R W_2 \xi \\ &\quad + \tfrac{1-\alpha}{\alpha}\xi^T W_1^T R W_1 \xi + \tfrac{\alpha}{1-\alpha}\xi^T W_2^T R W_2 \xi.\end{aligned}$$

By noting that $\frac{1-\alpha}{\alpha}$ and its inverse are positive, one can rewrite the last term of the previous equation as follows

$$\Theta(\alpha, R) = \xi^T W_1^T R W_1 \xi + \xi^T W_2^T R W_2 \xi - 2\xi^T W_1^T R W_2 \xi$$
$$+ \left(\sqrt{\tfrac{1-\alpha}{\alpha}} W_1 \xi + \sqrt{\tfrac{\alpha}{1-\alpha}} W_2 \xi\right)^T R \times \left(\sqrt{\tfrac{1-\alpha}{\alpha}} W_1 \xi + \sqrt{\tfrac{\alpha}{1-\alpha}} W_2 \xi\right).$$

Since the matrix R is positive definite, the last term of the previous expression is positive definite. Then it yields

$$\Theta(\alpha, R) \geq \xi^T W_1^T R W_1 \xi + \xi^T W_2^T R W_2 \xi - 2\xi^T W_1^T R W_2 \xi,$$

which concludes the proof. □

Remark 3 Obviously, our proposed new inequality is more conservative than the inequality from Lemma 4. Nevertheless, this advantage should be balanced with the numerical burden generated by the slack variable X as we will see with the numerical simulations.

In the sequel, these two lemmas will be employed in order to derive stability conditions for linear systems with time-varying delay. In particular a comparison on their use and on their induced conservatism will be discussed.

11.3 Stability Analysis of Time-Varying Delay Systems

We present in this sub-section a first stability result for time delay systems, which is based on the use of Jensen's inequality. Consider a linear time-delay system of the form:

$$\begin{cases} \dot{x}(t) = Ax(t) + A_d x(t - h(t)), & \forall t \geq 0, \\ x(t) = \phi(t), & \forall t \in [-h, 0], \end{cases} \tag{11.7}$$

where $x(t) \in \mathbb{R}^n$ is the state vector, ϕ is the initial condition and $A, A_d, A_D \in \mathbb{R}^{n\times n}$ are constant matrices. The delay is assumed to be time-varying. The classical constraints on the delay function are

$$\begin{array}{l} h(t) \in [0,\ h_M] \\ \dot{h}(t) \in (-\infty,\ d_M] \end{array}, \quad \forall t \geq 0, \tag{11.8}$$

for some scalars $0 \leq h_M$ and $d_M \leq 1$.

Based on the previous inequality and classical results on Lyapunov-Krasovskii functionals, three stability theorems are provided. The two first results are based on the use of the Jensen's inequality and Lemmas 4 and 5. The last one explicitly use the new inequality together with the reciprocally convex combination lemma provided in [15].

11.3.1 Stability Analysis Based on Jensen's Inequality

The stability analysis of the time-delay system (11.7) are established is the following theorems.

Theorem 1 *Assume that there exist two scalars $h_M > 0$, $d_M > 0$ and some matrices P, S, R, Q in $\mathbb{S}_+^n$ and X in $\mathbb{R}^{n\times n}$ such that the following LMIs are satisfied*

$$\Theta_0 = \begin{bmatrix} R & X \\ * & R \end{bmatrix} \succ 0,$$

$$\Psi_1(h_M, d_M) = \Psi_0(d_M) - \tfrac{1}{h_M}\Pi_0^T \Theta_0 \Pi_0 \prec 0,$$

where

$$\Psi_0(d_M) = F_1^T P F_0 + F_0^T P F_1 + \hat{S} + \hat{Q}(d_M) + h_M F_0^T R F_0, \tag{11.9}$$

with

$$\begin{aligned} \hat{Q}(d_M) &= diag(Q, -(1-d_M)Q, 0), \\ \hat{S} &= diag(S, 0, -S), \\ \Pi_0 &= \begin{bmatrix} F_2 \\ F_3 \end{bmatrix}, \end{aligned}$$

and

$$\begin{array}{l} F_0 = \begin{bmatrix} A \; A_d \; 0 \end{bmatrix}, \; F_1 = \begin{bmatrix} I \; 0 \; 0 \end{bmatrix}, \\ F_2 = \begin{bmatrix} I \; -I \; 0 \end{bmatrix}, \; F_3 = \begin{bmatrix} 0 \; I \; -I \end{bmatrix}. \end{array} \tag{11.10}$$

Then the system (11.7) *is asymptotically stable for the time-varying delay h satisfying* (11.8).

Notice that the last Theorem 1 does not depend on the lower bound of $\dot{h}(t)$.

Remark 4 It has to be noticed that Theorem 1 is similar to the one provided in [15] without lower bound of the delay and with $d_M = 1$.

The second theorem is formulated as follows:

Theorem 2 *Assume that there exist two scalars $h_M > 0$, $d_M > 0$ and some matrices P, S, R, Q in $\mathbb{S}_+^n$, such that the following LMI is satisfied*

$$\Psi_2(h_M, d_M) = \Psi_0(d_M) - \frac{1}{h_M}\Pi_1^T R \Pi_1 \prec 0, \tag{11.11}$$

where $\Psi_0(d_M)$ is defined in (11.9) *and where $\Pi_1 = F_2 - F_3$. Then the system* (11.7) *is asymptotically stable for the time-varying delay h satisfying* (11.8).

Proof Consider the Lyapunov functional given by

$$\begin{aligned} V(h, x_t, \dot{x}_t) = x^T(t)Px(t) &+ \int_{t-h_M}^{t} x^T(s)Sx(s)\mathrm{d}s \\ &+ \int_{t-h(t)}^{t} x^T(s)Qx(s)\mathrm{d}s \\ &+ \int_{t-h_M}^{t} (h_M - t + s)\dot{x}^T(s)R\dot{x}(s)\mathrm{d}s. \end{aligned} \quad (11.12)$$

This functional is positive definite since the matrices P, S, Q and R are symmetric positive definite. It has to be noticed that the functional is very simple compared to some functionals that can be found in the literature. Differentiating the functional (11.12) along the trajectories of (11.7) leads to:

$$\dot{V}(h, x_t, \dot{x}_t) = \xi^T(t)\Psi_0(\dot{h})\xi(t) - \int_{t-h_M}^{t} \dot{x}^T(s)R\dot{x}(s)\mathrm{d}s, \quad (11.13)$$

where $\xi(t) = \left[x^T(t)\ x^T(t-h(t))\ x^T(t-h_M)\right]^T$. This last equation has been obtained by noting that

$$x(t) = F_1\xi(t), \quad \dot{x}(t) = F_0\xi(t).$$

Splitting the integral into two integrals defined over the intervals $[t-h(t),\ t]$ and $[t-h_M,\ t-h(t)]$ and applying the Jensen's inequality (11.4), we get that

$$\begin{aligned} &\int_{t-h_M}^{t} \dot{x}^T(s)R\dot{x}(s)\mathrm{d}s \geq \\ &\quad \frac{1}{h(t)}\left[x(t) - x(t-h(t))\right]^T R\left[x(t) - x(t-h(t))\right] \\ &\quad \frac{1}{h_M - h(t)}\left[x(t-h(t)) - x(t-h_M)\right]^T R \\ &\quad \times \left[x(t-h(t)) - x(t-h_M))\right]. \end{aligned}$$

Rewriting the previous expression using the vector $\xi_0(t)$ and the matrices F_2 and F_3 yields to

$$\int_{t-h(t)}^{t} \dot{x}^T(s)R\dot{x}(s)\mathrm{d}s \geq \xi^T(t)\frac{F_2^T R F_2}{h(t)}\xi(t).$$

Applying the same method to the second integral, it yields to

$$\dot{V}(x_t, \dot{x}_t) \leq \xi^T(t)\left[\Phi_0(\dot{h}) - \frac{F_2^T R F_2}{h(t)} - \frac{F_3^T R F_3}{h_M - h(t)}\right]\xi(t).$$

The right hand side of the previous inequality is of the appropriate form to apply Lemma 4 and Lemma 5 which leads to the stability conditions given in Theorems 1 and 2, respectively. □

11.3.2 Stability Analysis Based on the New Integral Inequality

In this section, the assumptions on the delay function h are modified. In the following, the also classical constraints on the delay function becomes

$$\begin{array}{l} h(t) \in [h_m,\ h_M] \\ \dot{h}(t) \in [d_m,\ d_M] \end{array}, \quad \forall t \geq 0, \tag{11.14}$$

for some scalars $0 \leq h_m \leq h_M$ and $d_m \leq d_M \leq 1$.

Based on the previous inequality and classical results on Lyapunov-Krasovskii functionals, the stability theorem is provided.

Theorem 3 *Assume that there exist matrices P in $\mathbb{S}_+^{3n}$, S, Q, R in $\mathbb{S}_+^n$, and a matrix X in $\mathbb{R}^{2n\times 2n}$. Such that the following LMIs are satisfied for h in $\{h_m,\ h_M\}$ and for $\dot{h}$ in $\{d_m,\ d_M\}$*

$$\begin{array}{rl} \Theta_2 & = \begin{bmatrix} \tilde{R} & X \\ * & \tilde{R} \end{bmatrix} \succ 0, \\ \Phi(h, \dot{h}) & = \Phi_0(h, \dot{h}) - \frac{1}{h_M}\Gamma^T \Theta_2 \Gamma \prec 0, \end{array} \tag{11.15}$$

where

$$\begin{aligned} \Psi_0(h, \dot{h}) &= G_1^T(h) P G_0(\dot{h}) + G_0^T(\dot{h}) P G_1(h) + \hat{S} \\ &\quad + \hat{Q}(\dot{h}) + h_M G_0^T(\dot{h}) \hat{R} G_0(\dot{h}), \end{aligned}$$

$$\begin{aligned} \hat{Q}(\dot{h}) &= diag(Q, -(1-\dot{h})Q, 0_{3n}), \\ \hat{S} &= diag(S, 0, -S, 0_{2n}), \\ \hat{R} &= diag(R, 0_{3n}), \quad \tilde{R} = diag(R, R), \end{aligned}$$

and

$$\begin{array}{l} G_0(\dot{h}) = \begin{bmatrix} A & A_d & 0 & 0 & 0 \\ I & -(1-\dot{h})I & 0 & 0 & 0 \\ 0 & (1-\dot{h})I & -I & 0 & 0 \end{bmatrix}, \quad G_1(h) = \begin{bmatrix} I & 0 & 0 & 0 & 0 \\ 0 & 0 & 0 & hI & 0 \\ 0 & 0 & 0 & 0 & (h_M - h)I \end{bmatrix}, \\ \Gamma = \begin{bmatrix} G_2^T & G_3^T & G_4^T & G_5^T \end{bmatrix}^T, \\ G_2 = \begin{bmatrix} I & -I & 0 & 0 & 0 \end{bmatrix}, \qquad G_3 = \sqrt{3}\begin{bmatrix} I & I & 0 & -2I & 0 \end{bmatrix}, \\ G_4 = \begin{bmatrix} 0 & -I & I & 0 & 0 \end{bmatrix}, \qquad G_5 = \sqrt{3}\begin{bmatrix} 0 & I & I & 0 & -2I \end{bmatrix}. \end{array} \tag{11.16}$$

Then the system (11.7) *is asymptotically stable for the time-varying delay h satisfying* (11.14).

Proof Consider the Lyapunov-Krasovskii functional given by

$$\begin{aligned} V(h, x_t, \dot{x}_t) &= \zeta_0^T(t)P\zeta_0(t) + \int_{t-h(t)}^t x^T(s)Qx(s)\mathrm{d}s \\ &+ \int_{t-h_M}^t x^T(s)Sx(s)\mathrm{d}s \\ &+ \int_{t-h_M}^t (h_M - t + s)\dot{x}^T(s)R\dot{x}(s)\mathrm{d}s, \end{aligned} \tag{11.17}$$

where

$$\zeta_0(t) = \begin{bmatrix} x(t) \\ \int_{t-h(t)}^t x(s)\mathrm{d}s \\ \int_{t-h_M}^{t-h(t)} x(s)\mathrm{d}s \end{bmatrix}.$$

This functional is positive definite since the matrices P, S and R are symmetric positive definite. This functional is simple compared to some functionals that can be found in the literature. The only novelty remains in the introduction of the signals $\int_{t-h(t)}^t x(s)\mathrm{d}s$ and $\int_{t-h_M}^{t-h(t)} x(s)\mathrm{d}s$ in the functional which recall the method originally proposed by [5]. Differentiating the functional (11.17) along the trajectories of (11.7) leads to:

$$\dot{V}(h, x_t, \dot{x}_t) = \zeta^T(t)\Phi_0(h)\zeta(t) - \int_{t-h_M}^{t} \dot{x}^T(s)R\dot{x}(s)\mathrm{d}s, \tag{11.18}$$

where

$$\zeta(t) = \begin{bmatrix} x(t) \\ x(t-h(t)) \\ x(t-h_M) \\ \frac{1}{h(t)}\int_{t-h(t)}^t x(s)\mathrm{d}s \\ \frac{1}{h_M-h(t)}\int_{t-h_M}^{t-h(t)} x(s)\mathrm{d}s \end{bmatrix}.$$

This equation has been obtained by noting that

$$\zeta_0(t) = G_1(h)\zeta(t), \quad \dot{\zeta}(t) = G_0(\dot{h})\zeta(t).$$

Following the same procedure as in Theorem 2, the use of Lemma 1 leads to

$$\begin{aligned} \int_{t-h(t)}^t \dot{x}^T(s)R\dot{x}(s)\mathrm{d}s \geq & \frac{1}{h(t)}\zeta^T(t)\,[x(t) - x(t-h(t))]^T R\,[x(t) - x(t-h(t))] \\ & \frac{12}{h(t)}\left[\frac{x(t)+x(t-h(t))}{2} - \frac{1}{h(t)}\int_{t-h(t)}^t x(s)\mathrm{d}s\right]^T R \\ & \times \left[\frac{x(t)+x(t-h(t))}{2} - \frac{1}{h(t)}\int_{t-h(t)}^t x(s)\mathrm{d}s\right]. \end{aligned}$$

Then applying again Lemma 1 to the second integral and rewriting the previous expression using the vector $\zeta(t)$ and the matrices G_i for $i = 2, \dots 5$ yields to

$$\int_{t-h_M}^{t} \dot{x}^T(s)R\dot{x}(s)\mathrm{d}s \geq \zeta^T(t)\frac{(G_2^T RG_2+G_3^T RG_3)}{h(t)}\zeta(t) + \zeta^T(t)\frac{(G_4^T RG_4+G_5^T RG_5)}{h_m-h(t)}\zeta(t).$$

Applying the same method to the second integral and replacing the integral in (11.18) by its bounds, it yields

$$\dot{V}(x_t, \dot{x}_t) \leq \zeta^T(t)\left[\Phi_0(h, \dot{h}) - \frac{(G_2^T RG_2+G_3^T RG_3)}{h(t)} - \frac{(G_4^T RG_4+G_5^T RG_5)}{h_m-h(t)}\right]\zeta(t).$$

The right hand side of the previous inequality is exactly of the form described in Lemma 5. According to Lemma 4, the derivative of the Lyapunov Krasovskii functional is negative definite if

$$\Phi(h, \dot{h}) < 0, \forall (h, \dot{h}(t)) \in [0, h_M] \times [d_m, d_M]. \tag{11.19}$$

As the matrix $\Phi(h, \dot{h})$ is linear with respect to $h(t)$ and $\dot{h}(t)$, a sufficient condition is to test the condition on its vertices, which concludes the proof. the conditions of Theorem 3 are satisfied. □

Remark 5 The Lyapunov-Krasovskii functionals employed to derive the stability condition of Theorem 3 is very simple compared to the one that can be found in the literature (see for instance [18, 20] among many others). However, the use of the inequality proposed in Lemma 1 leads to a considerable reduction of the conservatism. This will be shown in the example section.

Remark 6 Any results based on the new inequality proposed by Lemma 1 combined with the reciprocal convexity result of Lemma 5 will be provided in the present chapter. Indeed, the combination of these two lemmas does not give efficient stability conditions compared to Theorem 3.

11.4 Example

Let consider the following example coming from the dynamics of machining chatter [19]:

$$\begin{aligned} \dot{x}(t) &= Ax(t) + Bu(t), \\ y(t) &= Cx(t) \end{aligned}$$

with

$$A = \begin{bmatrix} 0 & 0 & 1 & 0 \\ 0 & 0 & 0 & 1 \\ -10 & 10 & 0 & 0 \\ 5 & -15 & 0 & -0.25 \end{bmatrix}, \quad B = \begin{bmatrix} 0 \\ 0 \\ 1 \\ 0 \end{bmatrix}, \quad C = \begin{bmatrix} 1 & 0 & 0 & 0 \end{bmatrix}.$$

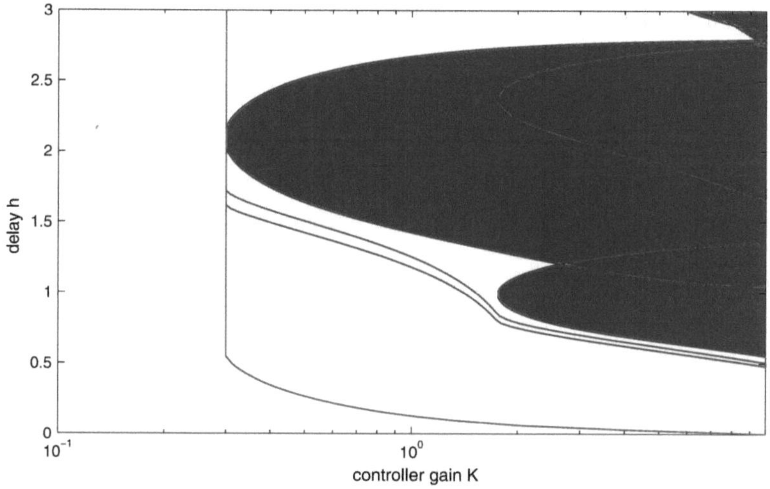

Fig. 11.1 Stability regions in the (K, h) plane

A delayed static output feedback controller is proposed:

$$u(t) = -Ky(t) + Ky(t-h),$$

where K is the gain of the controller and h is an unknown constant delay. The resulting dynamics is thus modeled by a time-delay system:

$$\dot{x}(t) = A_0 x(t) + A_1 x(t-h),$$

with $A_0 = A - BCK$ and $A_1 = BCK$. Applying theorem 3, we obtain the Fig. 11.1, the stability regions in the (K, h) plane. The blue region represents the instability region which have been calculated using a griding over K along with the `allmargin` function of the Control Toolbox of Matlab©. Using theorem 3, we have obtained inner approximations of the stability region delimited by colored curves. The red curve corresponds to the maximal allowable delay when Jensen's Lemma is used when establishing the stability criterion. In that case, we have used the results of [6]. The maximal allowable delay found by [17] which uses the first version of Wirtinger's Lemma is depicted over the green curve. Finally, the magenta curve corresponds to the maximal allowable delay when the improved Wirtinger's Lemma (i.e. theorem 3) is used when establishing the stability criterion. Clearly, theorem 3 improves drastically the delay upperbound compared to classical methods of the literature even if in that case, it not possible to detect a delay range stability pocket excluding a zero delay.

11.5 Conclusions

This chapter presents a new integral inequality which has been proved to be suitable for the stability analysis of time-delay systems. Combining it with the reciprocal convexity lemma leads to less conservative stability conditions than the usual ones which consider the Lyapunov-Krasovskii approach. This approach has been then employed to to study the stabilization of a regenerative chatter of machines tools.

Acknowledgments This work was supported by ANR project LimICoS contract number 12 BS03 005 01.

References

1. Ariba, Y., Gouaisbaut, F.: An augmented model for robust stability analysis of time-varying delay systems. Int. J. Control **82**, 1616–1626 (2009)
2. Ariba Y., Gouaisbaut F., and Johansson K.H. (2010) Stability interval for time-varying delay systems. In: 49th IEEE Conference on Decision and Control (CDC), pp. 1017–1022
3. Briat, C.: Convergence and equivalence results for the Jensen's inequality - application to time-delay and sampled-data systems. IEEE Trans. Autom. Control **56**(7), 1660–1665 (2011)
4. Fujioka, H.: Stability analysis of systems with aperiodic sample-and-hold devices. Autom. **45**(3), 771–775 (2009)
5. Gu K., Kharitonov V.-L., Chen J.: Stability of time-delay systems. Birkhauser, Boston (2003)
6. He, Y., Wang, Q.G., Xie, L., Lin, C.: Further improvement of free-weighting matrices technique for systems with time-varying delay. IEEE Trans. Automat. Control **52**, 293–299 (2007)
7. Jiang, X., Han, Q.L.: Delay-dependent robust stability for uncertain linear systems with interval time-varying delay. Automatica **42**(6), 1059–1065 (2006)
8. Kao, C.Y., Rantzer, A.: Stability analysis of systems with uncertain time-varying delays. Automatica **43**(6), 959–970 (2007)
9. Kim, J.H.: Note on stability of linear systems with time-varying delay. Automatica **47**(9), 2118–2121 (2011)
10. Liu, K., Fridman, E.: Wirtinger's inequality and Lyapunov-based sampled-data stabilization. Automatica **48**(1), 102–108 (2012)
11. Manitius A.Z., Olbrot A.W.: Finite specturm assignment problem for systems with delays. IEEE Trans. Autom. Control, **AC-24**(4), 541–553 (1979)
12. Mirkin, L.: Some remarks on the use of time-varying delay to model sample-and-hold circuits. IEEE Trans. Autom. Control **52**(6), 1009–1112 (2007)
13. Moon, Y.S., Park, P., Kwon, W.H., Lee, Y.S.: Delay-dependent robust stabilization of uncertain state-delayed systems. Int. J. Control **74**(14), 1447–1455 (2001)
14. Niculescu, S.-I.: Delay effects on stability: A Robust Control Approach. Springer, Heidelberg (2001)
15. Park, P.G., Ko, J.W., Jeong, C.: Reciprocally convex approach to stability of systems with time-varying delays. Automatica **47**, 235–238 (2011)
16. Seuret, A., Gouaisbaut, F.: Wirtinger-based integral inequality: application to time-delay systems. Automatica **49**(9), 2860–2866 (2013)
17. Seuret A., Gouaisbaut F.: On the use of the wirtinger's inequalities for time-delay systems. In: Proceedings of the 10th IFAC workshop on time delay systems (IFAC TDS'12), Boston, MA, USA (2012)
18. Shao, H.: New delay-dependent stability criteria for systems with interval delay. Automatica **45**(3), 744–749 (2009)

19. Sipahi, R., Niculescu, S.-I., Abdallah, C.T., Michiels, W., Gu, K.: Stability and stabilization of systems with time delay. IEEE Control Syst. **31**(1), 38–65 (2011)
20. Sun, J., Liu, G.P., Chen, J., Rees, D.: Improved delay-range-dependent stability criteria for linear systems with time-varying delays. Automatica **46**(2), 466–470 (2010)

Part III
Applications

Chapter 12
Control-Oriented Input-Delay Model of the Distributed Temperature of a SI Engine Exhaust Catalyst

Delphine Bresch-Pietri, Thomas Leroy and Nicolas Petit

Abstract This chapter aims at showing how a particular class of input delay ordinary differential equations, in which the time- and input-dependent delay is defined through an implicit integral equation, can be used to model accurately the internal temperature of a Spark-Ignited engine catalyst. The modeling approach is grounded on a one-dimensional distributed parameter model, which is approximated by a time-varying first-order delay system whose dynamics parameters (time constant, delay, gains) are obtained through a simple analytic reduction procedure. Following recent works, the distributed heat generation resulting from pollutant conversion is shown here to be equivalent to an inlet temperature entering the system at a virtual front inside the catalyst. The gain of this new input introduces a coupling to account for the conversion efficiency. Relevance of this real-time compliant model is qualitatively supported by experimental data.

12.1 Introduction

Elements of context. Automotive Spark-Ignited (SI) engines are equipped with a Three-Way Catalyst (TWC) located in the exhaust line. This after-treatment device aims at reducing the three major pollutants resulting from the combustion: hydrocar-

D. Bresch-Pietri (✉)
Department of Mechanical Engineering, Massachusetts Institute of Technology,
77 Massachusetts Avenue, Cambridge, MA 02139, USA
e-mail: dbp@mit.edu

T. Leroy
Département Contrôle, Signal et Système in IFP Energies nouvelles,
1-4 Av. du Bois Préau, 92852 Rueil Malmaison, France
e-mail: thomas.leroy@ifpen.fr

N. Petit
Centre Automatique et Systèmes, Unité Mathématiques et Systèmes
at MINES ParisTech, 60 Bd St Michel, 75272 Paris, France
e-mail: nicolas.petit@mines-paristech.fr

A. Seuret et al. (eds.), *Low-Complexity Controllers for Time-Delay Systems*,
Advances in Delays and Dynamics 2, DOI: 10.1007/978-3-319-05576-3_12,

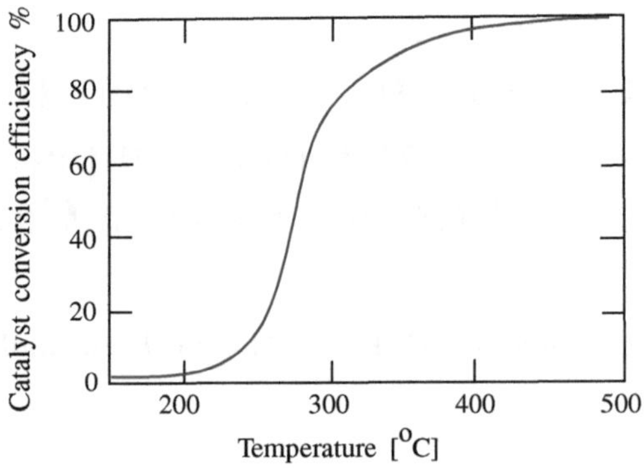

Fig. 12.1 Conversion efficiency (jointly for CO, HC and NO_x) as a function of temperature for typical catalytic converter (*Source* [9])

bons HC, carbon monoxide CO and nitrogen oxide NO_x. Yet, conversion efficiency highly depends on the catalyst temperature[1] [9, 10], as presented in Fig. 12.1. Right after a cold start of the engine, temperatures are too low to activate chemical reactions and the catalyst conversion ratio is poor [18]. Therefore, speed-up of the catalyst warm-up is a point of critical importance to reach high level of pollutant conversion.

Classically, warm-up strategies are performed by increasing the exhaust gas temperatures via combustion timing shifting [8]. This open-loop technique leads to a faster heating of the catalyst but also yields combustion efficiency degradation and therefore substantial consumption increase. This increase must be limited to its strict minimum. For this reason, it is of prime importance to determine when the catalyst has reached its light-off temperature[2] to obtain a satisfactory compromise between pollutant emissions and consumption. When this light-off temperature is obtained, standard combustion can be performed and the consumption can simply go back to a standard level.

Motivations for real-time wall temperature modeling. Unfortunately, no temperature sensor is commercially embedded to provide an information on the distributed wall temperature. In commercial line products, determination of the switch time is currently achieved from the measurements provided by a commercially embedded temperature sensor located into the cooling system. Indeed, the thermal behavior of the water cooling system can be indirectly related to the engine and exhaust line temperatures. Yet, this information is highly uncertain and has no reason to be repeatable (which cannot be allowed anymore as driving cycles tend do diversify).

An alternative is to rely on models. Unfortunately, catalyst temperature models that have been proposed in the literature are either mean-value (spatially lumped) model [11], which do not take into account the inherent distributed nature of the catalyst and can therefore reveal highly inaccurate, or Partial Differential Equations(PDE) modeling [5, 12, 15] with complex representations of the heat release by chemical

[1] It also highly depends on the Air/Fuel Ratio, the influence of which is not considered here.

[2] Defined here as the temperature at which the catalyst becomes more than 90 % effective.

reactions, which give very accurate estimation of the light-off temperature but are discarded from real-time implementations by the induced computational burden.

Contribution and organization of the chapter. In this chapter, we propose to use a semi-lumped model of these PDE equations. Following the overture presented in [13], we show how a first-order input-delay dynamics relates the inlet gas temperature to a punctual wall catalyst temperature. The obtained model belongs to a particular class of time- and input-dependent delay systems in which the delay is defined through an implicit integral equation which is representative of transport phenomena [17, 21, 23]. The chemical reactions inside the catalyst are simply represented as a fictitious second temperature front entering the catalyst afar off the physical catalyst inlet. This model is shown to be quite accurate, and of gentle implementation complexity.

The model presented here can be seen as a generalization of [13] to SI engines applications. The main modifications consists in the introduction of the catalyst conversion efficiency impacting the heat release. This efficiency depends on the output of the model, resulting into an additional coupling which does not tamper with the stability of the model.

This chapter is organized as follows. Section 12.2 presents the catalyst under consideration in experiments. In Sect. 12.3, we detail the PDE temperature modeling which is used in Sect. 12.4 to derive a first-order input-delay model through analytic formula stemming from simple operational calculus. Relevance of the proposed model is discussed at the light of simulations performed on experimental data. We conclude with directions of future works such as prediction-based control strategy.

12.2 Experimental Set-Up

The catalyst under consideration in this study is mounted at the outlet of a 2L four-cylinder turbocharged SI engine, downstream the turbine. Figure 12.2 presents a scheme of the catalyst under consideration. It is composed of two separated monoliths [19] which, in the following analysis, for the sake of clarity, are not distinguished. For experimental studies and comparisons, the catalyst has been instrumented with two internal temperature sensors. Such sensors located are not embedded inside any commercial line product.

Figure 12.3 presents experimental results obtained at test bench during a New European Driving Cycle (NEDC) cycle. Histories of both the exhaust mass flow and the temperature located upstream the catalyst are reported in Fig. 12.3a. These quantities are the inputs of the model proposed in this chapter. The exhaust mass flow is a fast-varying variable closely related to the engine torque output. In Fig. 12.3b, both monolith temperatures of Fig. 12.2 are given for a cycle without warm-up strategy. By comparing these two curves between them and against the inlet gas temperature, one can notice the very low-pass filter role of the catalyst (see the signals T_w^1 and T_w^2 on Fig. 12.3b). We will account for this in our model simplification.

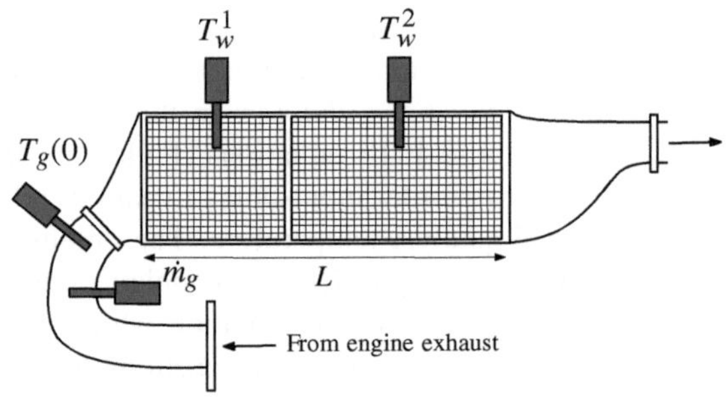

Fig. 12.2 Experimental catalyst composed of two monoliths. Two sensors permit to measure the wall temperature in the center of each monolith. Test-bench is also equipped with inlet temperature and mass flow sensors.

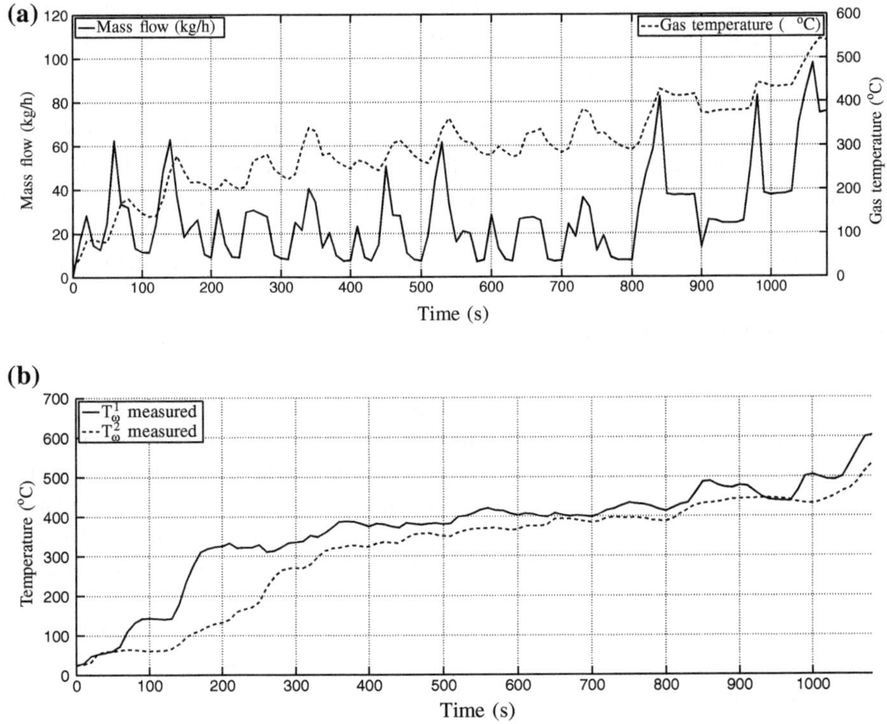

Fig. 12.3 Experimental results on European driving cycle (NEDC). **a** Exhaust gas flow and temperature. **b** First and second monolith temperature.

12.3 Partial Differential Equation Model

We now refer to Fig. 12.4, where a schematic representation of the monolith is given. Exhaust burned gas enter the monolith at $x = 0$ and convective exchange with the wall occur all along the monolith, i.e. for $x = 0$ to $x = L$, yielding to distributed temperature profiles of the gas $T_g(x, t)$ and the catalyst wall $T_w(x, t)$ (as underlined in [20, 22], the axial conduction can be neglected).

We consider the following coupled linear infinite dimensional thermal dynamics

$$\begin{cases} \dfrac{\partial T_w}{\partial t}(x, t) = k_1(T_g(x, t) - T_w(x, t)) + \Psi(x, t, T_w(x, t)) & (12.1) \\ \dot{m}_g \dfrac{\partial T_g}{\partial x}(x, t) = k_2(T_w(x, t) - T_g(x, t)) & (12.2) \end{cases}$$

where ψ is a distributed time-varying source term, related to the chemical reaction occurring inside the catalyst and the constants $k_1, k_2 > 0$ are defined as

$$k_1 = \frac{h_I P_I}{A_w \rho_w C p_w}, \quad k_2 = \frac{h_I P_I}{C p_g}.$$

Notations are gathered in Table 12.1 in Appendix. Such a model is considered for example in [12]. It encompasses the detailed modeling (12.14)–(12.15) given in Appendix, provided that a few simplifications are performed:

- conduction ($\lambda_w \partial^2 T_g / \partial x^2$) into the monolith is neglected compared to convection exchanges;
- gas storage is considered as very small compared to the monolith one, i.e. $\rho_g C p_g << \rho_w C p_w$;
- convective exchanges with the atmosphere are neglected compared to the one with the exhaust gas (this is assumed on for the sake of simplicity of the exposition and can easily be relaxed).

The source term ψ gathers the sum of the enthalpies of the various reactions taking place inside the catalyst. It can be effectively represented as

$$\Psi(x, t, T_w) = \begin{cases} \psi(x, t, T_w) & \text{for} \quad 0 \leq x \leq L_r, \\ 0 & \text{for} \quad L_r < x \leq L, \end{cases}$$

where L_r is the length of the portion of the catalyst where the heat is released. This model is illustrated in Fig. 12.4. Note that the source term also depends on the wall temperature. This is important to study the light-off process: for moderate temperature, the conversion efficiency highly depends on the wall temperature and this dependence cannot reasonably be neglected.

In this chapter, we propose to represent the conversion efficiency of the catalyst as a function of a punctual wall temperature, at a position L_η (potentially varying with

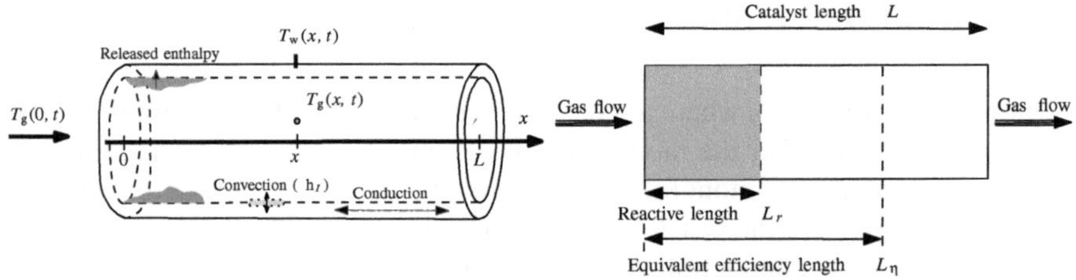

Fig. 12.4 Schematic views of the distributed profile temperature inside a catalyst jointly with thermal exchanges (*left*) and of the proposed modeling inspired from [13] (*right*). The conversion is assumed to take place on an upstream part of the catalyst of length L_r. The temperature used to determine the catalyst efficiency is located at length $L_\eta \leq L$

Table 12.1 Notations

Symbol	Description	Unit
A_w	Catalyst wall area	m^2
A_g	Catalyst efficient area	m^2
Cp_w	Catalyst wall heat capacity	$J/kg/m^3$
Cp_g	Gas heat capacity	$J/kg/m^3$
D	Model delay	s
ΔH_{HC}	Unity enthalpy from HC conversion	J/mol
ΔH_{CO}	Unity enthalpy from CO conversion	J/mol
ΔH_{NOx}	Unity enthalpy from NO_x conversion	J/mol
h_I	Internal convection coefficient	$J/K/m^2$
h_O	External convection coefficient	$J/K/m^2$
h_i	Enthalpy of the *i*th species	J
λ_w	Wall conduction coefficient	$J/K/m^2$
$\dot{m}_g$	Gas mass flow rate	kg/s
P_I	Internal catalyst perimeter	m
P_O	External catalyst perimeter	m
R_i	Reaction rate of the *i*th species	–
ρ_w	Catalyst wall volumetric mass	kg/m^3
T_w	Distributed monolith temperature	K
T_w^1	Wall temperature in the middle of the first monolith	K
T_w^2	Wall temperature in the middle of the second monolith	K
T_g	Distributed gas temperature	K
τ	Model time constant	s

aging). Experimental determination of this efficiency was performed and follows the tendency of Fig. 12.1. In the following, it is called η, considered as a known function, and we focus on the design of a simple model of the wall temperature at L_η , handling the potential variability of this position.

12.4 Approaching the Dynamics by an Input-Delay Ordinary Differential Equation

12.4.1 Operational Calculus without Source Term

Before detailing the global model that we propose to use, we focus on the analysis of the "purely thermal" behavior of the PDEmodel, i.e. without any source term. This case is representative of a temperature interval below 250–300, where chemical conversion is almost ineffective.

Claim. Assume $\psi = 0$. In the range of low (time domain) frequencies, the distributed parameter model 12.1–12.2 can be approximated by the following set of first-order delayed equations

$$\forall 0 \leq x \leq L, \quad \tau(x,t)\frac{dT_w(x,t)}{dt} = -T_w(x,t) + T_g(0, t - D(x,t)) \quad (12.3)$$

with

$$\tau(x,t) = \frac{1}{k_1} + \nu\delta(x,t), \quad D(x,t) = (1-\nu)\delta(x,t), \quad (12.4)$$

where ν is a given constant in $[0, 1]$ and δ is defined through the integral equation

$$\int_{t-\delta(x,t)}^{t} \frac{k_1}{k_2}\dot{m}_g(s)ds = x. \quad (12.5)$$

The relation (12.5) implicitly defines a transport delay through past values of the gas flow rate. It corresponds to a transport phenomenon occurring over a length x with a speed $\frac{k_1}{k_2}\dot{m}_g$ accordingly to a Plug-Flow assumption [17]. This time can be understood as a residence time into the monolith (see [4]). As the two main effects of the gas residence inside the monolith are transport and exchange with the monolith, it can reasonably be separated into a first order dynamics with a pure delay effect. The tuning parameter ν can be determined via dedicated tests (e.g. given operating points, for which the engine is initially cold and requested torque and engine speed are kept constant) and allows this model to qualitatively represent a relatively vast range of catalyst devices.

12.4.1.1 Formulation of Claim 12.4.1

Transport delay

By taking a spatial derivative of 12.1, a time-derivative of 12.2 and matching terms with 12.1–12.2, one can obtain the decoupled equations, for all $x \in [0, L]$,

$$\begin{cases} \dot{m}_g(t)\dfrac{\partial^2 T_w}{\partial x \partial t} = -k_2\dfrac{\partial T_w}{\partial t} - k_1\dot{m}_g(t)\dfrac{\partial T_w}{\partial x}, \\ \dot{m}_g(t)\dfrac{\partial^2 T_g}{\partial x \partial t} + \ddot{m}_g(t)\dfrac{\partial T_g}{\partial x} = -k_2\dfrac{\partial T_g}{\partial t} - k_1\dot{m}_g(t)\dfrac{\partial T_g}{\partial x}, \end{cases}$$

where the first equation defining T_w can be reformulated using a spatial Laplace transform (operational calculus) to get

$$\forall t \geq 0, \quad (\dot{m}_g(t)p + k_2)\frac{d\hat{T}_w}{dt} = -k_1\dot{m}_g(t)p\hat{T}_w(p, t).$$

This scalar system can be solved as

$$\hat{T}_w(p, t) = \exp\left(-\left[\int_{t_0}^{t} \frac{k_1\dot{m}_g(s)p}{\dot{m}_g(s)p + k_2}ds\right]\right)\hat{T}_w(p, t_0),$$

where t_0 is such that $t_0 \leq t$. The catalyst, as is visible from experimental data reported in Fig. 12.3b, is relatively non-sensitive to high-frequencies. Consequently, by considering only low-level spatial frequencies (i.e., $\dot{m}_g p << k_2$ for any gas flow $\dot{m}_g$), the term below the integral can be substantially simplified. Rewriting the resulting equation into the usual space domain gives

$$\forall x \in [0, L], \quad T_w(x, t) = T_w\left(x - \left[\int_{t_0}^{t} \frac{k_1}{k_2}\dot{m}_s(s)ds\right], t_0\right).$$

Formally, one can define $\delta(x, t) \geq 0$ such that

$$x - \left[\int_{t-\delta(x,t)}^{t} \frac{k_1}{k_2}\dot{m}_s(s)ds\right] = 0,$$

which is equivalent to the implicit integral equation (12.5). Consequently, the wall temperature at abscissa x is formally delayed by

$$\forall x \in [0, L], \quad T_w(x, t) = T_w(0, t - \delta(x, t)). \tag{12.6}$$

First-order model
From there, it is possible to relate the dynamics under consideration to the gas inlet temperature. Consider for a moment that $\delta(x)$ is constant with respect to time. Then, writing (12.6) in the time Laplace domain, jointly with 12.1 for $x = 0$, one directly obtains for all $x \in [0, L]$

$$\hat{T}_w(x, s) = k_1 \frac{e^{-\delta(x)s}}{s + k_1} \hat{T}_g(0, s). \tag{12.7}$$

Finally, following the same steps as previously, it is possible to only consider low frequencies ($s << 1$). By following the elements presented in [16],

$$e^{-\delta(x)s} \approx \frac{e^{-(1-\nu)\delta(x)s}}{\nu\delta(x)s + 1},$$

with a constant $\nu \in [0, 1]$, (12.7) rewrites for low frequencies as

$$\hat{T}_w(x, s) = \frac{e^{-(1-\nu)\delta(x)s}}{\left(\frac{1}{k_1} + \nu\delta(x)\right) s + 1} \hat{T}_g(x, 0).$$

By formally generalizing this relation to a time-varying residence time $\delta(x, t)$, one obtains the dynamics formulated in Claim 12.4.1.

12.4.1.2 Validation of the Reduced Model (12.3–12.5) Using Experimental Data

To illustrate Claim 12.4.1, simulation results of the temperature inside the wall catalyst at two different locations are pictured in Fig. 12.5. The two simulation results have been obtained respectively with the distributed parameter model 12.1–12.2 (with $\psi = 0$, i.e. neglecting the enthalpy flows) and with the proposed simplified dynamics (12.3)–(12.5). The inputs used for the two models (gas mass flow rate and gas inlet temperature) are data recorded during a NEDC cycle. They are pictured in Fig. 12.3a.

Implementation of the proposed model was performed with a forward Euler approximation of (12.3). The value of δ is determined based on a trapezoidal approximation of (12.5) with a simple calculation procedure: the integral left-hand side of (12.5) is an increasing function of δ, equal to zero for $\delta = 0$; therefore, we calculate it recursively for increasing values of δ, starting with $\delta = 0$, stopping when reaching or passing x.

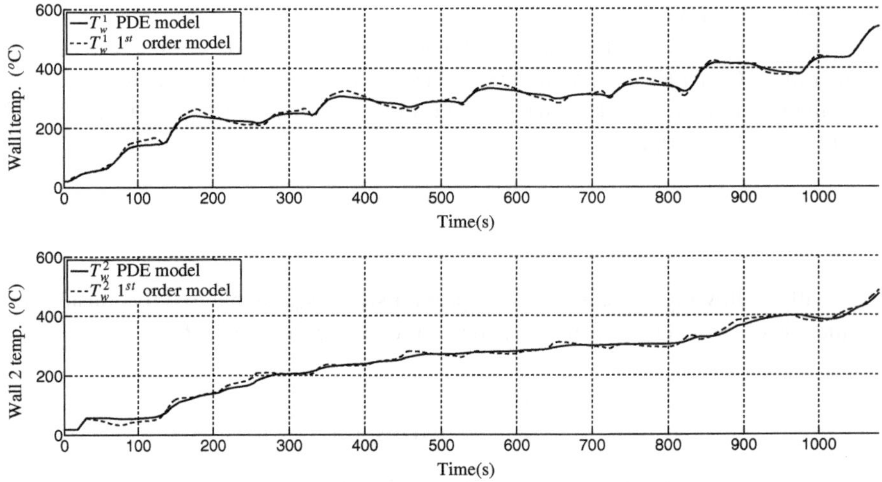

Fig. 12.5 Simulation comparison between the temperature representation of the model 12.1–12.2 with $\psi = 0$ and the reduced first-order input-delay model (12.3)–(12.5). The inputs of the model are NEDC variations pictured in Fig. 12.3a

The simulated temperature in Fig. 12.5 almost perfectly matches the one computed with the PDE model. As these performances are obtained for very demanding external conditions (large gas mass flow rate variations), one can reasonably expect similar behavior on different kinds of driving conditions. For typically encountered input signals, the PDE model is well represented by the model of Claim 12.4.1.

Nevertheless, these models cannot completely match experimentally measured data, as the heat released from chemical reactions is neglected. We now investigate this point.

12.4.2 Including Chemical Reactions Energy

To account for the source term ψ, we propose to consider the pollutant conversion effects as a second temperature front T_{eq} occurring at virtual position L_r inside the catalyst.[3] This model allows one to exploit the linearity of the dynamics (12.3)–(12.5), through a superposition principle, to distinguish the $T_g(0)$ effects from the pollutant conversion effect. This model approach is pictured on Fig. 12.6.

---

[3] In details, this fictitious length does not exactly match the physical non-reactive length introduced earlier in Sect. 12.3. Yet, for sake of simplicity, we assume here that they are identical.

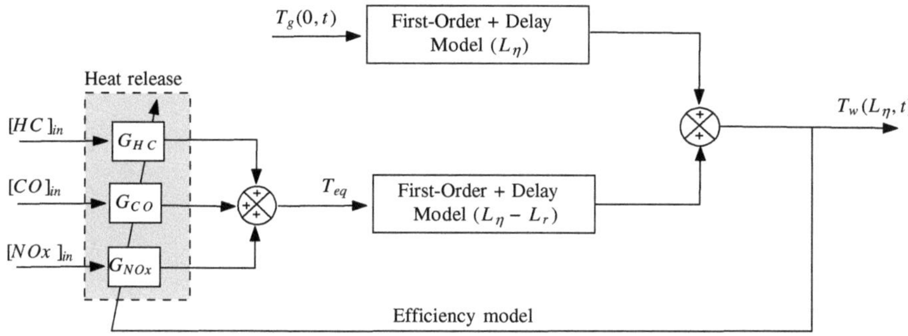

Fig. 12.6 Proposed catalyst temperature model (12.9)–(12.13). The pollutant conversion effects (HC, CO and NO_x) are assimilated to a front of temperature T_{eq} propagating on a virtual length $L_\eta - L_r$, while the gas heating occurs on the complete length L_η. The model is also fed by the gas mass flow rate $\dot{m}_g$ which is not represented here for sake of clarity

For steady-state conditions, energy balance for the system can be written as

$$\dot{m}_g Cp_g \underbrace{(T_g(0) - T_g(L))}_{\triangleq T_{eq}} + \eta(T_w(L_\eta)) \sum_{i=1}^{N} \Delta H_i [x_i]_{in} = 0,$$

where $[x_i]_{in}$ are the inlet pollutant concentrations. Typically, three main pollutants are considered ($N = 3$), i.e. hydrocarbons (HC), carbon monoxide (CO) and nitrogen oxides (NO_x). They result in three steady-state gains

$$G_{HC} = \eta(T_w(L_\eta)) \frac{\Delta H_{HC}}{\dot{m}_g Cp_g}, \quad G_{CO} = \eta(T_w(L_\eta)) \frac{\Delta H_{CO}}{\dot{m}_g Cp_g}$$
$$\text{and} \quad G_{NO_x} = \eta(T_w(L_\eta)) \frac{\Delta H_{NO_x}}{\dot{m}_g Cp_g},$$

where the unity enthalpy ΔH_{HC}, ΔH_{CO} and ΔH_{NO_x} are known constants. These gains are then used to calculate an equivalent temperature

$$T_{eq} = G_{HC}[HC]_{in} + G_{CO}[CO]_{in} + G_{NO_x}[NO_x]_{in}. \tag{12.8}$$

In practice, the pollutant concentrations are not measured but can be effectively estimated, e.g. by look-up tables.

An important point to notice is the appearance of the temperature at length L_η as a parametrization of the conversion efficiency. This yields a coupling represented in Fig. 12.6 under a closed-loop form.

We summarize this approach by the following claim.

Claim. For any source term ψ, the wall catalyst temperature at position L_η can be efficiently represented as

$$T_w(L_\eta) = T_w^{th} + T_w^{\psi}, \tag{12.9}$$

where T_w^{th} satisfies

$$\tau(L_\eta, t)\frac{dT_w^{th}}{dt} = -\, T_w^{th}(t) + T_g(0, t - D(L_\eta, t)), \tag{12.10}$$

and T_w^{ψ} satisfies

$$\tau(L_\eta - L_r, t)\frac{dT_w^{\psi}}{dt} = -\, T_w^{\psi}(t) + T_{eq}(0, t - D(L_\eta - L_r, t)). \tag{12.11}$$

T_{eq} is defined in (12.8), the time constant τ and the delay D are defined for $x \in [0, L]$ as

$$\tau(x, t) = \frac{1}{k_1} + \nu\delta(x, t)\,, \quad D(x, t) = (1 - \nu)\delta(x, t), \tag{12.12}$$

with ν a given constant in $[0, 1]$ and δ defined through the integral equation

$$\int_{t-\delta(x,t)}^{t} \frac{k_1}{k_2}\dot{m}_g(s)ds = x. \tag{12.13}$$

It is worth noticing that the catalyst temperature at any position $x \in [0, L]$ can also be computed by a similar procedure, provided one has value of the steady-state gains (correspondingly, $T_w(L_\eta, .)$ has to be calculated independently).

12.4.3 Validation of the Proposed Model on Experimental Data

Simulation results of the wall catalyst temperature at two positions (described in Fig. 12.2) are provided in Fig. 12.7 and compared to experimental measurements. These measurements were obtained on a NEDC cycle, with an initially cold catalyst. The tuning parameter is set to $\nu = 0.4$.

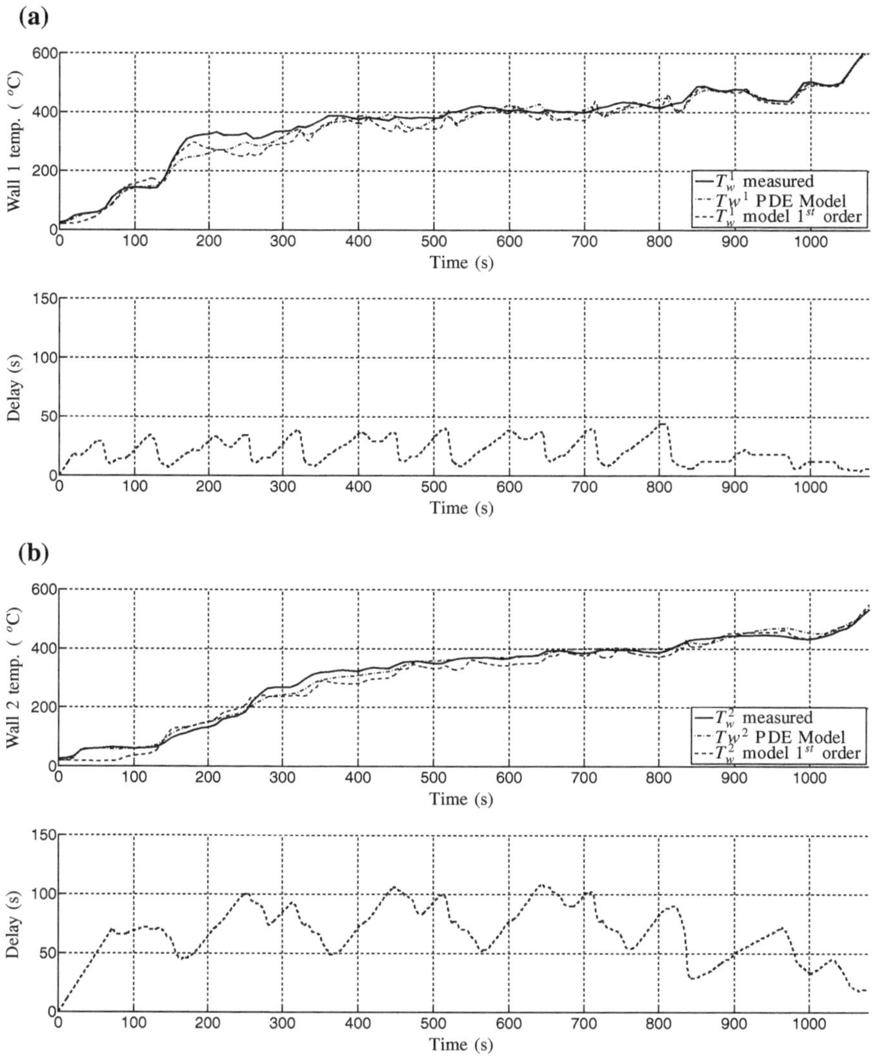

Fig. 12.7 Comparison between the proposed model and experimental data at two location inside the monolith. **a** Simulation results for the first-order input-delay model approach, compared to experimental data in the center of the first monolith, and the corresponding, delay used in the model. **b** Simulation results for the first-order input-delay model approach, compared to experimental data in the center of the second monolith, and the corresponding delay used in the model.

One can easily notice that the computed temperatures catch both short-term and long-term variations of the true signals. As previously, it is worth noticing that the inputs corresponding to this NEDC cycle are highly variable and therefore this test case is challenging.

The proposed model is simple enough to be implemented in real-time and provides accurate estimation of the wall catalyst. One clear advantage of the proposed technique that is worth noticing is that it provides insight into the temperature everywhere inside the monolith. A lumped model (or 0D-model) like the one presented in [11] for example, cannot achieve this. Also worth noticing is the fact that aging of the catalyst can be accounted for by updating L_η.

To feed the model, values for various inputs, presented in Fig.12.6, are necessary: the mass flow rate (already modeled for cylinder charge estimation and combustion control), the inlet gas temperature (the modeling of which has been widely investigated in the literature, see [6, 7]) and the pollutant emissions upstream of the catalyst.

12.5 Conclusion

In this chapter, a simple infinite dimensional model of the wall catalyst has been presented and a corresponding first-order input-delay reduction has been performed. Besides its real-time compliance, this model also exhibits the interesting property of providing an estimation of the wall temperature without requiring additional offline tuning procedures.

This reduced input-delay model is of particular interest to design strategies which, compared to the existing ones, enable to detect light-off independently on the driving cycle (see [2] for further experimental-based simulation illustrating this point). Other control strategies exploiting the delay representation such as the prediction-based techniques proposed in [3] can also be considered. Further evaluation of their potential merits is the scope of future work. Another direction of work is the evaluation of the robustness of this model to input estimation errors (in particular with respect to inlet pollutant emissions which are not measured but obtained through look-up tables).

Appendix

In this appendix, we provide a more detailed modeling of the thermal exchanges occurring in the catalyst, from which 12.1–12.2 is a simplification. Following [5], a thermal balance of the gas leads to the equation

$$\rho_g A_g C p_g \frac{\partial T_g}{\partial t} + \dot{m}_g C p_g \frac{\partial T_g}{\partial x} = h_I P_I (T_w(x,t) - T_g(x,t)), \tag{12.14}$$

where the first term on the left hand side accounts for the gas energy storage, the second one for transport and the right-hand term for convective exchanges. A similar balance for the wall yields

$$\rho_w A_w C p_w \frac{\partial T_w}{\partial t} = \lambda_w \frac{\partial^2 T_w}{\partial x^2} + \sum_{i=1}^{N} R_i h_i + h_I P_I (T_g(x,t) - T_w(x,t)) + h_O P_O (T_{amb} - T_w(x,t)), \quad (12.15)$$

where the left-hand side still accounts for the energy storage and the right-hand side represents respectively: (i) the conduction/diffusion inside the monolith; (ii) the enthalpy flow of the N chemical reactions occurring inside the catalyst (mainly, $N = 3$); (iii) the exchange respectively with the gas and the atmosphere.

One can notice that, following [13], no transport occurs in the (solid) wall. In more details, a mass balance of the species in presence can be established.

The species concentrations inside the monolith are necessary to determine the reaction terms R_i in the enthalpy flows. Two additional equations per species are also necessary (one for the gas and one for the monolith, see [1, 14]).

References

1. Benjamin, S.F., Roberts, C.A.: Automotive catalyst warm-up to light-off by pulsating engine exhaust. Int. J. Engine Res. **5**(2), 125–147 (2004)
2. Bresch-Pietri, D., Leroy, T., Petit, N.: Control-oriented time-varying input-delayed temperature model for SI engine exhaust catalyst. In: Proceedings of the American Control Conference (2013)
3. Bresch-Pietri, D., Leroy, T., Petit, N.: Estimation of the distributed temperature of a SI engine catalyst for light-off strategy. In: Proceedings of the European Control Conference (2013)
4. Danckwerts, P.V.: Continuous flow systems: distribution of residence times. Chem. Eng. Sci. **2**(1), 1–13 (1953)
5. Depcik, C., Assanis, D.: One-dimensional automotive catalyst modeling. Prog. Energy Combust. Sci. **31**(4), 308–369 (2005)
6. Eriksson, L.: Mean value models for exhaust system temperatures. In: Proceedings of the Society of Automotive Engineering World Congress, number 2002–01-0374 (2005)
7. Fu, H., Chen, X., Shilling, I., Richardson, S.: A one-dimensional model for heat transfer in engine exhaust systems. In: Proceedings of the Society of Automotive Engineering World Congress, number 2005–01-0696 (2005)
8. Guzzella, L., Onder, C.H.: Introduction to Modeling and Control of Internal Combustion Engine Systems. Springer, New York (2010)
9. Heywood, J.B.: Internal Combustion Engine Fundamentals. McGraw-Hill, New York (1988)
10. Kiencke, U., Nielsen, L.: Automotive Control Systems. Springer, Berlin (2000)
11. Kum, D., Peng, H., Bucknor, N.K.: Optimal energy and catalyst temperature management of plug-in hybrid electric vehicles for minimum fuel consumption and tail-pipe emissions. IEEE Trans. Control Syst. Technol. **21**(99), 14–26 (1999)
12. Laing, P.M., Shane, M.D., Son, S., Adamczyk, A.A., Li, P.: A simplified approach to modeling exhaust system emissions: SIMTWC. SAE paper, 01–3476 (1999)
13. Lepreux, O.: Model-based temperature control of a diesel oxidation catalyst. Ph.D. thesis, MINES Paristech (2010)
14. Olsson, L., Andersson, B.: Kinetic modelling in automotive catalysis. Top. Catal. **28**(1), 89–98 (2004)
15. Onorati, A., D'Errico, G., Ferrari, G.: 1D fluid dynamic modeling of unsteady reacting flows in the exhaust system with catalytic converter for SI engines. SAE Trans. **109**(2000–01–0210), 89–103 (2000)

16. Pekar, L., Kureckova, E.: Rational approximations for time-delay systems: case studies. In: Proceedings of the Conference on Mathematical and computational Methods in Science and Engineering, pp. 217–222 (2011)
17. Perry, R.H., Green, D.W., Maloney, J.O.: Perry's Chemical Engineers' Handbook, vol. 7. McGraw-Hill, New York (1984)
18. Please, C.P., Hagan, P.S., Schwendeman, D.W.: Light-off behavior of catalytic converters. SIAM J. Appl. Math. **54**(1), 72–92 (1994)
19. Ramanathan, K., West, D.H., Balakotaiah, V.: Optimal design of catalytic converters for minimizing cold-start emissions. Catal. Today **98**(3), 357–373 (2004)
20. Vardi, J., Biller, W.F.: Thermal behavior of exhaust gas catalytic convertor. Ind. Eng. Chem. Process Des. Dev. **7**(1), 83–90 (1968)
21. Wu, M., Wang, C., Cao, W., Lai, X., Chen, X.: Design and application of generalized predictive control strategy with closed-loop identification for burn-through point in sintering process. Control Eng. Pract. **20**, 1065–1074 (2012)
22. Young, L.C., Finlayson, B.A.: Mathematical models of the monolith catalytic converter. AIChE J. **22**(2), 343–353 (1976)
23. Zenger, K., Ylinen, R.: Simulation of variable delays in material transport models. Math. Comput. Simul. **37**, 52–72 (1994)

Chapter 13
Suppressing Stick-Slip Oscillations in Oilwell Drillstrings

Belem Saldivar, Sabine Mondié and Alexandre Seuret

Abstract In drilling operations, the drillstring interaction with the borehole gives rise to a wide variety of non-desired oscillations. The main types of drilling vibrations are torsional (stick-slip), axial (bit-bounce) and lateral (whirling). The analysis and modeling of rotary drilling vibrations is a topic whose economical interest has been renewed by recent oil fields discoveries leading to a growing literature. This contribution addresses the problem of modeling and control of the most critical vibration mode: the stick-slip phenomenon. The wave equation subject to mixed boundary conditions is used to reproduce torsional oscillations of the drillstring. By means of a direct transformation we derive an input-output model described by a neutral-type time-delay equation which clearly simplifies the system simulation. A dissipativity analysis of the wave equation model leads to the design of a stabilizing controller guaranteeing a non-growth of energy during the drilling process. Furthermore, by means of a polytopic representation of the neutral-type time-delay model, stability and stabilization conditions in terms of Linear Matrix Inequalities (LMIs) are given. The performance of the proposed approaches is highlighted through simulations showing an effective suppression of the stick-slip phenomenon.

B. Saldivar (✉) · S. Mondié
Centro de Investigación y de Estudios Avanzados del Instituto Politécnico Nacional (CINVESTAV-IPN), Av. Instituto Politécnico Nacional No. 2508, Col San Pedro Zacatenco, C.P. 07360 México, D.F., Mexico
e-mail: msaldivar@ctrl.cinvestav.mx

S. Mondié
e-mail: smondie@ctrl.cinvestav.mx

A. Seuret
Laboratoire d'Analyse et d'Architecture des Systèmes (LAAS-CNRS), 7 Avenue du Colonel Roche, 31400 Toulouse, France
e-mail: alexandre.seuret@laas.fr

A. Seuret et al. (eds.), *Low-Complexity Controllers for Time-Delay Systems*, Advances in Delays and Dynamics 2, DOI: 10.1007/978-3-319-05576-3_13,

13.1 Introduction

The presence of drillstring vibrations is the main cause of loss of performance in the perforation process for oil and gas. It provokes premature wear and tear of drilling equipment resulting in fatigue and induced failures such as pipe wash-out and twist-off [14]. It also causes significant wastage of drilling energy [13] and may induce wellbore instabilities reducing the directional control and its overall shape [7]. In the oil industry, the improvement of drilling performance is a matter of crucial economical interest.

Downhole measurements show that applying a constant rotary speed at the surface does not necessarily translate into a steady rotational motion of the bit. In fact, the downhole torsional speed typically exhibits large amplitude fluctuations during a significant fraction of the drilling time. This self-excited rotational motion, also known as stick-slip, is induced by the nonlinear relationship between the torque and the angular velocity at the bit [12]. The torsional flexibility of the drilling assembly exacerbates a non-uniform oscillatory behavior causing rotational speeds as high as ten times the nominal rotary table speed or a total standstill of the bit [20].

Even though new technology has been deployed, such phenomena still occur affecting considerably on drilling costs and daily operations. Before 1960s, studies to treat the vibration problem were focused on material strength of the drillstring components but the trends have since changed to emphasize on its dynamic behavior [12]. Some classical control techniques for suppressing the stick-slip phenomenon are: the introduction of a soft torque rotary system at the top of the drillstring [18], where the underlying idea is making the top rotary system behave in a soft manner rather than as a fixed heavy flywheel so that the torsional waves arriving at the surface are absorbed breaking the harmful cycling motion; the introduction of a vibration absorber at the top of the drillstring [12]; the introduction of a PID controller at the surface in order to control the rotary speed and the introduction of additional friction at the bit [16]. A more sophisticated control methodology is used in [21] where a linear $\mathcal{H}_\infty$ control is used to suppress stick-slip motion at the bit.

This work is concerned with the problem of modeling and control of torsional vibrations in a vertical oilwell drilling system. The wave equation subject to mixed boundary conditions allows to reproduce the rotational motion of the drillstring. Through the D'Alembert transformation we are able to obtain an input-output representation of the system described by a nonlinear neutral-type time-delay model involving only the variables of our interest. Furthermore we deal with the system stabilization based on both proposed modeling strategies by using Lyapunov theory. A proposal of energy function allows a dissipativity analysis of the drilling system leading to the design of a control law guaranteeing the suppression of torsional oscillations. Finally, based on a polytopic representation of the neutral-type time-delay model, stability and stabilization conditions in terms of LMIs are given. This approach leads the design of a nonlinear controller ensuring the exponential stability of the closed loop system and consequently the elimination of the stick-slip.

13.2 Drilling System Modeling

The main process during well drilling for oil is the creation of borehole by a rock-cutting tool called bit. The drillstring consists of the Bottom Hole Assembly (BHA) and drillpipes screwed end to end to each other to form a long pipe. The BHA comprises the bit, stabilizers to prevent the drillstring from unbalancing, and pipe sections relatively heavy known as drill collars. The drillstring is rotated from the surface by an electrical motor located at the rotary table.

Drilling system complexity brings out a modelling and a control problem. The model must accurately describe the most important phenomena arising in real wells and it has to be simple enough for analysis and control purposes.

The wave equation is widely used to reproduce the oscillatory behavior of physical systems. In the 1960s, some analyses have been carried out using the classical wave equation to describe the torsional behavior of drillstring assemblies [8]. This modeling strategy was also adopted in [6] and [22].

By means of a direct transformation, an input-output model described by a neutral-type time-delay equation, which clearly simplifies the analysis and simulations, is obtained. The procedure allowing transforming the partial differential equation (PDE) model to a delay system of neutral type was presented for the first time in [1], nowadays the method is addressed in a large number of contributions; some of them are cited below. A linear delay equation arising from the wave equation that describes the torsional behavior of a flexible rod with a mass is presented in [9]. In [2, 5, 17], drilling vibrations analyses were performed based on a neutral type time delay model derived from the wave equation model. Similar reductions of analogous PDE boundary value problems to delay equations have relevance to a broad range of other physical, biological and engineering problems, e.g. transmission line oscillators [3, 4].

13.2.1 Distributed Parameter Model

The purely torsional excitations of a drill-string of length L, described by the rotary angle $\theta(\xi, t)$ can be modeled by the wave equation [5, 6, 11, 22]:

$$\frac{\partial^2\theta}{\partial\xi^2}(\xi,t) = p^2\frac{\partial^2\theta}{\partial t^2}(\xi,t), \quad p=\sqrt{I/GJ}, \quad \xi\in[0,L], \quad t>0, \tag{13.1}$$

coupled to the boundary conditions:

$$GJ\frac{\partial\theta}{\partial\xi}(0,t) = c_{\mathrm{a}}\left(\frac{\partial\theta}{\partial t}(0,t)-\Omega(t)\right); \tag{13.2}$$

$$GJ\frac{\partial\theta}{\partial\xi}(L,t) + I_{\mathrm{B}}\frac{\partial^2\theta}{\partial t^2}(L,t) = -T\left(\frac{\partial\theta}{\partial t}(L,t)\right),$$

where $\Omega(t)$ is the angular velocity coming from the rotor that does not match the rotational speed of the load $\frac{\partial \theta}{\partial t}(0, t)$. This sliding speed results in the local torsion of the drillstring. The parameters I, G and J are the inertia, the shear modulus and the geometrical moment of inertia respectively, I_{B} is a lumped inertia that represent the assembly at the bottom hole. The undriven extremity ($\xi = L$) is subject to a torque T which is a function of the bit speed $\frac{\partial \theta}{\partial t}(L, t)$. The nature of $T\left(\frac{\partial \theta}{\partial t}(L, t)\right)$ is a critical aspect of the model and is discussed in details below.

13.2.2 Neutral-Type Time-Delay Model

By reducing the boundary value problem (13.1–13.2) to a neutral-type time delay equation we are able to exploit techniques from dynamic systems theory to gain insight into the complexity involved in the analysis and simulation of infinite dimensional systems.

The general solution of the one-dimensional wave equation that models vibrations of a string can be written as: $\theta(\xi, t) = \phi(t+p\xi)+\psi(t-p\xi)$, where the variables $\gamma = t+p\xi$, and $\eta = t-p\xi$ result from the application of the method of characteristics to (13.1), ϕ and ψ are continuously differentiable real-valued functions of one variable. The boundary conditions can be rewritten as:

$$p\frac{\partial \phi}{\partial \gamma}(t) - p\frac{\partial \psi}{\partial \eta}(t) = \frac{c_a}{GJ}\left(\frac{\partial \phi}{\partial \gamma}(t) + \frac{\partial \psi}{\partial \eta}(t) - \Omega(t)\right),$$

$$\begin{aligned} p\left(\frac{\partial \phi}{\partial \gamma}(t+\Gamma) - \frac{\partial \psi}{\partial \eta}(t-\Gamma)\right) =& -\frac{I_B}{GJ}\frac{\partial^2 \phi}{\partial \gamma^2}(t+\Gamma) - \frac{I_B}{GJ}\frac{\partial^2 \psi}{\partial \eta^2}(t-\Gamma) \\ & - \frac{1}{GJ}T\left(\frac{\partial \phi}{\partial \gamma}(t+\Gamma) + \frac{\partial \psi}{\partial \eta}(t-\Gamma)\right), \end{aligned}$$

where $\Gamma = pL$. Defining $z(t)$ as the angular velocity at the bottom extremity of the drill string: $z(t) = \frac{\partial \theta(L,t)}{\partial t} = \dot{\phi}(t+\tau) + \dot{\psi}(t-\tau)$, we obtain, under appropriate manipulations, the following neutral-type time-delay equation which describes the input-output behavior of the system [17]:

$$\begin{aligned} \dot{z}(t) - \Upsilon\dot{z}(t-2\Gamma) =& -\Psi z(t) - \Upsilon\Psi z(t-2\Gamma) - \frac{1}{I_{\mathrm{B}}}T\left(z(t)\right) \\ & + \frac{1}{I_{\mathrm{B}}}\Upsilon T\left(z(t-2\Gamma)\right) + \Pi\Omega(t-\Gamma), \end{aligned} \tag{13.3}$$

where $z(t)$ is the angular velocity at the bottom end, and

$$\Pi = \frac{2\Psi c_{\mathrm{a}}}{c_{\mathrm{a}} + \sqrt{IGJ}} \quad \Upsilon = \frac{c_{\mathrm{a}} - \sqrt{IGJ}}{c_{\mathrm{a}} + \sqrt{IGJ}}, \quad \Psi = \frac{\sqrt{IGJ}}{I_{\mathrm{B}}}, \quad \Gamma = \sqrt{\frac{I}{GJ}}L.$$

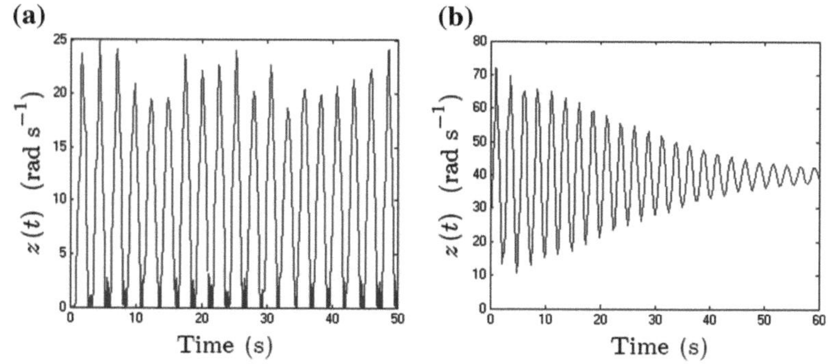

Fig. 13.1 Angular velocity at the bottom extremity $z(t)$ for: **a** $\Omega(t) = 10\,\mathrm{rad\,s^{-1}}$ (stick-slip), **b** $\Omega(t) = 40\,\mathrm{rad\,s^{-1}}$

It should be pointed out that this is an exact transformation, i.e., the model (13.3) is not an approximation of the wave equation model.

13.2.3 Stick-Slip Phenomenon

It is usually assumed that the growth of instabilities leading eventually to stick-slip oscillations arises from the friction model, which empirically captures the bit-rock interaction. The most common friction models include velocity weakening laws [6], stiction plus Coulomb friction [12, 21], and the Stribeck effect characterized by a decreasing friction-velocity map localized around zero velocity [15].

The following nonlinear expression, introduced in [15], allows to approximate the physical phenomenon at the bit level:

$$T\left(z(t)\right) = c_{\mathrm{b}} z(t) + W_{\mathrm{ob}} R_{\mathrm{b}} \mu_{\mathrm{b}}\left(z(t)\right) \mathrm{sign}\left(z(t)\right). \tag{13.4}$$

The first term (viscous damping torque) represents the influence of the fluid drilling and the second one (dry friction torque) approximates the bit-rock contact, R_{b} stands for the bit radius and W_{ob} is the weight on the bit. The bit dry friction coefficient $\mu_{\mathrm{b}}\left(z(t)\right)$ is described by

$$\mu_{\mathrm{b}}\left(z(t)\right) = \mu_{\mathrm{cb}} + (\mu_{\mathrm{sb}} - \mu_{\mathrm{cb}}) \mathrm{e}^{-\frac{\gamma_{\mathrm{b}}}{v_{\mathrm{f}}}|z(t)|}, \tag{13.5}$$

where $0 < \gamma_{\mathrm{b}} < 1$ is a constant defining the velocity decrease rate and $\mu_{\mathrm{cb}}, \mu_{\mathrm{sb}} \in (0, 1)$ are the Coulomb and static friction coefficients. The exponential decaying behavior of $T\left(z(t)\right)$ coincides with experimental torque values. Figure 13.1 shows the stick-slip behavior at the bit level. Notice that torsional vibrations are reduced by increasing the velocity $\Omega(t)$ provided by the rotary table. The model parameters used for the simulations presented in the sequel are [6, 15].

13.3 Control Design to Suppress the Stick-Slip Phenomenon

In this section, based on the different modeling approaches to describe the drilling dynamics, we introduce two control strategies to eliminate the stick-slip phenomenon. First, we deal with the distributed parameter model of the system; the proposal of an energy function allows to find a control law that ensures the energy dissipation during the drilling operation. Next, based on a polytopic representation of the nonlinear neutral-type time-delay model, we find an effective stabilizing controller to suppress the stick-slip. The angular velocity coming from the rotary table $\Omega(t)$ is taken as control input. We consider the structure $\Omega(t) = u(t) + \Omega_0$, where $u(t)$ is a stabilizing controller which should eliminate the stick-slip phenomenon and Ω_0 is a prescribed reference velocity.

13.3.1 Wave Equation Approach

The normalized model of the drilling system is:

$$z_{tt}(\sigma,t) = a z_{\sigma\sigma}(\sigma,t) - d z_t(\sigma,t), \quad t > 0, \quad 0 < \sigma < 1, \tag{13.6}$$

where $a = \frac{GJ}{IL^2}$ and $d = \frac{\beta}{I}$, coupled to the mixed boundary conditions:

$$\begin{aligned} z_\sigma(0,t) &= g\left(z_t(0,t) - u(t)\right), \quad \sigma \in (0,1), \quad t > 0, \\ z_\sigma(1,t) &= -k z_t(1,t) - q\mu_b(z_t(1,t))\text{sign}(z_t(1,t)) - h z_{tt}(1,t), \end{aligned} \tag{13.7}$$

where $g = \frac{c_a L}{GJ}$, $k = \frac{c_b L}{GJ}$, $q = \frac{W_{ob} R_b L}{GJ}$, $h = \frac{I_B L}{GJ}$.

Consider the energy function

$$E(t) = \int_0^1 a z_\sigma^2(\sigma,t)d\sigma + \int_0^1 z_t^2(\sigma,t)d\sigma + ahz_t^2(1,t). \tag{13.8}$$

Differentiating $E(t)$ along (13.6), yields

$$\begin{aligned} \frac{d}{dt}E(t) &= 2\int_0^1 a z_\sigma(\sigma,t) z_{t\sigma}(\sigma,t)d\sigma + 2\int_0^1 z_t(\sigma,t) z_{tt}(\sigma,t)d\sigma \\ &\quad + 2ahz_t(1,t)z_{tt}(1,t), \\ &= 2a\int_0^1 \left[z_\sigma(\sigma,t)z_{t\sigma}(\sigma,t) + z_t(\sigma,t)z_{\sigma\sigma}(\sigma,t)\right]d\sigma \\ &\quad - 2d\int_0^1 z_t(\sigma,t)z_t(\sigma,t)d\sigma + 2ahz_t(1,t)z_{tt}(1,t). \end{aligned}$$

Integrating by parts and substituting the boundary conditions (13.7) gives

$$\begin{aligned}\int_0^1 z_t(\sigma,t)z_{\sigma\sigma}(\sigma,t)d\sigma &= z_t(\sigma,t)z_\sigma(\sigma,t)|_0^1 - \int_0^1 z_\sigma(\sigma,t)z_{t\sigma}(\sigma,t)d\sigma\\ &= z_t(1,t)(-kz_t(1,t) - q\mu_b(z_t(1,t))\text{sign}(z_t(1,t)) - hz_{tt}(1,t))\\ &\quad - gz_t(0,t)\,(z_t(0,t)-u(t)) - \int_0^1 z_\sigma(\sigma,t)z_{t\sigma}(\sigma,t)d\sigma.\end{aligned}$$

Hence,

$$\begin{aligned}\frac{d}{dt}E(t) &= 2az_t(1,t)(-kz_t(1,t) - q\mu_b(z_t(1,t))\text{sign}(z_t(1,t)) - hz_{tt}(1,t))\\ &\quad + 2a\int_0^1 z_\sigma(\sigma,t)z_{t\sigma}(\sigma,t)d\sigma - 2agz_t(0,t)\,(z_t(0,t)-u(t))\\ &\quad - 2a\int_0^1 z_\sigma(\sigma,t)z_{t\sigma}(\sigma,t)d\sigma - 2d\int_0^1 z_t(\sigma,t)z_t(\sigma,t)d\sigma\\ &\quad + 2ahz_t(1,t)z_{tt}(1,t),\end{aligned}$$

since $\mu_b(z_t(1,t))\,\text{sign}(z_t(1,t))z_t(1,t) = \mu_b(z_t(1,t))\,|z_t(1,t)|$, we have

$$\begin{aligned}\frac{d}{dt}E(t) =& -2aq\mu_b(z_t(1,t))\,|z_t(1,t)| - 2d\int_0^1 z_t^2(\sigma,t)d\sigma\\ & - 2agz_t(0,t)\,(z_t(0,t)-u(t)) - 2akz_t^2(1,t).\end{aligned} \tag{13.9}$$

In order to ensure the dissipativity of the system, the control law $u(t)$ should allow the negativity of (13.9). Choosing as a control law the following expression:

$$u(t) = (1-c_1)z_t(0,t) + 2c_1z_t(1,t) - c_1\frac{(z_t(1,t))^2}{z_t(0,t)}, \tag{13.10}$$

where $c_1 > 0$ is a free design parameter we obtain that

$$\begin{aligned}\frac{d}{dt}E(t) =& -2aq\mu_b(z_t(1,t))\,|z_t(1,t)| - 2d\int_0^1 z_t^2(\zeta,t)d\zeta\\ & - 2agc_1\,(z_t(1,t)-z_t(0,t))^2 - 2akz_t^2(1,t).\end{aligned}$$

Taking into account that $\mu_b(z_t(1,t)) > 0$ and a, q, k, d and g are positive constants we find that $\frac{d}{dt}E(t) \le 0$. The energy dissipation of the drilling system is established:

Proposition 1 *For all solutions of (13.6) under the boundary conditions (13.7), the energy given by (13.8) does not grow if the control law (13.10) is applied.*

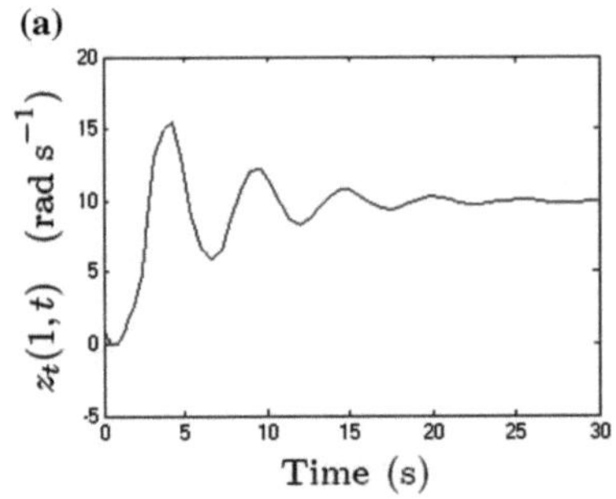

Fig. 13.2 Stick-slip elimination by means of applying the control law (13.10) for: **a** $\Omega_0 = 10$ rad$\dot{s}^{-1}$, **b** $\Omega_0 = 20$ rad s^{-1}

The simulation result of Fig. 13.2 shows that the stick-slip vibrations are reduced by means of the application of the control law (13.10) ensuring a dissipative behavior of the drilling system.

13.3.2 Polytopic Approximation Approach

We investigate the stabilization of a class of nonlinear neutral-type time-delay systems that can be transformed into a multi-model system, i.e., a set of linear models non-linearly weighted represented as follows:

$$\dot{x}(t) - D\dot{x}(t-\tau_1) = \sum\nolimits_{i\in I^r} h_i(x_t)\left\{A_i x(t) + A_{i\tau_1} x(t-\tau_1) + Bu(t)\right\}, \quad (13.11)$$

where $\tau_1 > 0$ is a constant time delay, the set I^r is the set of integers $\{1, \ldots, r\}$, r is the number of subsystems required to describe the multi-model system. The functions $h_i(\cdot)$ are scalar weighting functions satisfying the convexity conditions:

$$\sum\nolimits_{i\in I^r} h_i(x_t) = 1 \quad \forall i = 1, \ldots, r, \quad h_i(x_t) \geq 0. \quad (13.12)$$

First, we analyze the α-stability of the open loop system:

$$\dot{x}(t) - D\dot{x}(t-\tau_1) = \sum\nolimits_{i\in I^r} h_i(x_t)\left\{A_i x(t) + A_{i\tau_1} x(t-\tau_1)\right\}. \quad (13.13)$$

To guarantee that the difference operator is stable we assume $|D| < 1$. The change of variable $x_\alpha(t) = \mathrm{e}^{\alpha t} x(t)$ transforms the system (13.13) into:

$$\begin{aligned}\dot{x}_\alpha(t)D - \mathrm{e}^{\alpha\tau_1}\dot{x}_\alpha(t-\tau_1) = &\sum\nolimits_{i\in I^r} h_i(x_t)\left\{(A_i + \alpha I_n)\, x_\alpha(t)\right.\\ &\left.+\mathrm{e}^{\alpha\tau_1}\left(A_{i\tau_1} - \alpha D\right) x_\alpha(t-\tau_1)\right\}. \quad (13.14)\end{aligned}$$

The proposal is to find conditions for which the solution $x_\alpha = 0$ of the transformed system (13.14) is asymptotically stable. Clearly, these conditions will guarantee the exponential stability of system (13.13).

The following result is based on the ideas presented in [19] which addresses the stabilization problem for nonlinear retarded-type time-delay systems.

Theorem 1 *The solution $x(t) = 0$ of the system (13.13) is α-stable if there exist matrices $0 < P_1 = P_1^T$, P_2, P_3, $Q = Q^T$ and $R = R^T$, such that for all $i \in I^r$ the following LMI is satisfied*

$$\begin{bmatrix} \Psi_i & P^T \begin{bmatrix} 0 \\ e^{\alpha\tau_1}\left(A_{i\tau_1} - \alpha D\right) \end{bmatrix} & P^T \begin{bmatrix} 0 \\ e^{\alpha\tau_1} D \end{bmatrix} \\ * & -R/\tau_1 & 0 \\ * & * & -Q \end{bmatrix} < 0, \tag{13.15}$$

$$P := \begin{bmatrix} P_1 & 0 \\ P_2 & P_3 \end{bmatrix}, \; P_1 = P_1^T > 0, \; \Lambda_i := A_i + \alpha I_n + e^{\alpha\tau_1}\left(A_{i\tau_1} - \alpha D\right),$$

$$\Psi_i := P^T \begin{bmatrix} 0 & I_n \\ \Lambda_i & -I_n \end{bmatrix} + \begin{bmatrix} 0 & I_n \\ \Lambda_i & -I_n \end{bmatrix}^T P + \begin{bmatrix} 0 & 0 \\ 0 & \tau_1 R + Q \end{bmatrix}.$$

Proof According to the Leibniz formula $x_\alpha(t - \tau_1) = x_\alpha(t) - \int_{t-\tau_1}^{t} \dot{x}_\alpha(s)ds$, we can write the system (13.14) as

$$\dot{x}_\alpha(t) - De^{\alpha\tau_1}\dot{x}_\alpha(t - \tau_1) = \sum\nolimits_{i\in I^r} h_i(x_t)\{-e^{\alpha\tau_1}\left[A_{i\tau_1} - \alpha D\right] \int_{t-\tau_1}^{t} \dot{x}_\alpha(s)ds$$
$$\left[A_i + \alpha I_n + e^{\alpha\tau_1}\left(A_{i\tau_1} - \alpha D\right)\right] x_\alpha(t)\}.$$

Using the descriptor form introduced in [10] we have

$$\dot{x}_\alpha(t) = y(t),$$
$$y(t) = \sum\nolimits_{i\in I^r} h_i(x_t)\{De^{\alpha\tau_1} y(t - \tau_1) + \Lambda_i x_\alpha(t)$$
$$- e^{\alpha\tau_1}\left(A_{i\tau_1} - \alpha D\right) \int_{t-\tau_1}^{t} y(s)ds\}, \quad \Lambda_i = A_i + \alpha I_n + e^{\alpha\tau_1}\left(A_{i\tau_1} - \alpha D\right),$$

Then, we can write

$$E \begin{bmatrix} \dot{x}_\alpha(t) \\ \dot{y}(t) \end{bmatrix} = \begin{bmatrix} y(t) \\ \sum_{i\in I^r} h_i(x_t) \cdot \lambda \end{bmatrix}, \quad E = diag\{I_n, 0\},$$
$$\lambda = -y(t) + De^{\alpha\tau_1} y(t - \tau_1) + \Lambda_i x_\alpha(t) - e^{\alpha\tau_1}\left(A_{i\tau_1} - \alpha D\right) \int_{t-\tau_1}^{t} y(s)ds.$$

Following [10], we use the Lyapunov-Krasovskii functional

$$V_\alpha(t) = \begin{bmatrix} x_\alpha(t) \\ y(t) \end{bmatrix}^T EP \begin{bmatrix} x_\alpha(t) \\ y(t) \end{bmatrix} + \int_{t-\tau_1}^{t} y^T(s)Qy(s)ds + \int_{-\tau_1}^{0}\int_{t+\theta}^{t} y^T(s)Ry(s)dsd\theta, \quad (13.16)$$

where $P = \begin{bmatrix} P_1 & 0 \\ P_2 & P_3 \end{bmatrix}$, $P_1 = P_1^T > 0$, $R > 0$, $Q > 0$.

The functional $V_\alpha(t)$ is positive definite since

$$\left[x_\alpha^T(t)\; y^T(t)\right] EP \begin{bmatrix} x_\alpha(t) \\ y(t) \end{bmatrix} = x_\alpha^T(t)P_1x_\alpha(t).$$

Notice that $EP = P^TE$, taking the derivative in t of $V_\alpha(t)$ we obtain

$$\dot{V}_\alpha(t) = 2\left[x_\alpha^T(t)\; y^T(t)\right]P^T \begin{bmatrix} y(t) \\ \sum_{i\in I^r} h_i(x_t)\cdot\lambda \end{bmatrix} + \tau_1 y^T(t)Ry(t) - \int_{t-\tau_1}^{t} y^T(s)Ry(s)ds + y^T(t)Qy(t) - y^T(t-\tau_1)Qy(t-\tau_1).$$

Setting $\xi = \left[x_\alpha^T(t)\; y^T(t)\; y^T(t-\tau_1)\right]^T$ we can write

$$\dot{V}_\alpha(t) = \xi^T \begin{bmatrix} \tilde{\Psi}_i & P^T\begin{bmatrix} 0 \\ e^{\alpha\tau_1}D \end{bmatrix} \\ \left[0\; e^{\alpha\tau_1}D^T\right]P & -Q \end{bmatrix} \xi + \Xi - \int_{t-\tau_1}^{t} y^T(s)Ry(s)ds,$$

where

$$\Xi = -2\int_{t-\tau_1}^{t} \left[x_\alpha^T(t)\; y^T(t)\right]P^T \begin{bmatrix} 0 \\ e^{\alpha\tau_1}\left(A_{i\tau_1} - \alpha D\right) \end{bmatrix} y(s)ds,$$

$$\tilde{\Psi}_i = \sum_{i\in I^r} h_i(x_t)\left\{P^T\begin{bmatrix} 0 & I_n \\ \Lambda_i & -I_n \end{bmatrix} + \begin{bmatrix} 0 & I_n \\ \Lambda_i & -I_n \end{bmatrix}^T P + \begin{bmatrix} 0 & 0 \\ 0 & \tau_1 R + Q \end{bmatrix}\right\}.$$

In order to obtain an upper bound on Ξ, we use the following property. For all vectors $a, b \in \mathscr{R}^n$ and positive definite matrix $R\in\mathscr{R}^{n\times n}$, the inequality $\pm 2a^Tb \leq a^TR^{-1}a + b^TRb$ is satisfied. Then, the following inequality holds

$$\Xi \leq \left[x_\alpha^T(t)\; y^T(t)\right]P^T \begin{bmatrix} 0 \\ e^{\alpha\tau_1}\left(A_{i\tau_1} - \alpha D\right) \end{bmatrix} \tau_1 R^{-1} \left[0\; e^{\alpha\tau_1}\left(A_{i\tau_1} - \alpha D\right)^T\right] P \begin{bmatrix} x_\alpha(t) \\ y(t) \end{bmatrix} + \int_{t-\tau_1}^{t} y^T(s)Ry(s)ds.$$

In view of the above inequalities, we have

$$\dot{V}_\alpha(t) \leq \xi^T \begin{bmatrix} \tilde{\Psi}_i & P^T \begin{bmatrix} 0 \\ \mathrm{e}^{\alpha\tau_1} D \end{bmatrix} \\ \begin{bmatrix} 0 \ \mathrm{e}^{\alpha\tau_1} D^T \end{bmatrix} P & -Q \end{bmatrix} \xi + \begin{bmatrix} x_\alpha(t) \\ y(t) \end{bmatrix}^T P^T$$

$$\times \begin{bmatrix} 0 \\ \mathrm{e}^{\alpha\tau_1} \left(A_{i\tau_1} - \alpha D\right) \end{bmatrix} \left(\tau_1 R^{-1}\right) \begin{bmatrix} 0 \\ \mathrm{e}^{\alpha\tau_1} \left(A_{i\tau_1} - \alpha D\right) \end{bmatrix}^T P \begin{bmatrix} x_\alpha(t) \\ y(t) \end{bmatrix}.$$

Finally, using Schur complements, the system (13.13) is asymptotically stable if the LMI condition (13.15) is satisfied. □

Having determined the criteria for exponential stability for the open loop system (13.13), the next step is to define an algorithm that allows the synthesis of a gain K such that the feedback control law $u(t) = Kx(t - \tau_1)$ exponentially stabilizes the closed loop system

$$\dot{x}(t) - D\dot{x}(t - \tau_1) = \sum\nolimits_{i \in I^r} h_i(x_t) \left\{A_i x(t) + \left(A_{i\tau_1} + BK\right) x(t - \tau_1)\right\}, \quad (13.17)$$

with a guaranteed decay rate α.

By replacing the matrix $A_{i\tau_1}$ by the matrix $A_{i\tau_1} + BK$ in Theorem 1 we obtain that the solution $x(t) = 0$ of the system (13.17) is α-stable if there exist matrices $0 < P_1 = P_1^T$, P_2, P_3, $Q = Q^T$, $R = R^T$ such that for all $i \in I^r$ the following bilinear matrix inequality (BMI) is satisfied

$$\begin{bmatrix} \Psi_i & P^T \begin{bmatrix} 0 \\ \mathrm{e}^{\alpha\tau_1} \chi \end{bmatrix} & P^T \begin{bmatrix} 0 \\ \mathrm{e}^{\alpha\tau_1} D \end{bmatrix} \\ * & -R/\tau_1 & 0 \\ * & * & -Q \end{bmatrix} < 0, \quad (13.18)$$

where

$$P := \begin{bmatrix} P_1 & 0 \\ P_2 & P_3 \end{bmatrix}, \quad P_1 = P_1^T > 0,$$

$$\Psi_i := P^T \begin{bmatrix} 0 & I_n \\ \Lambda_i & -I_n \end{bmatrix} + \begin{bmatrix} 0 & I_n \\ \Lambda_i & -I_n \end{bmatrix}^T P + \begin{bmatrix} 0 & 0 \\ 0 & \tau_1 R + Q \end{bmatrix},$$

$$\Lambda_i := A_i + \alpha I_n + \mathrm{e}^{\alpha\tau_1} \left(A_{i\tau_1} + BK - \alpha D\right), \ \chi := \left(A_{i\tau_1} + BK - \alpha D\right).$$

A well known synthesis gain technique which overcome the bilinearity of the conditions consists in to set $P_3 = \varepsilon P_2$, $\varepsilon \in \mathscr{R}$, where P_2 is a nonsingular matrix, and $\bar{P} = P_2^{-1}$. Defining $\bar{P}_1 = \bar{P}^T P_1 \bar{P}$, $\bar{R} = \bar{P}^T R \bar{P}$, and $Y = K\bar{P}$ and multiplying the right side of (13.18) by $\Delta_3 = diag\left\{\bar{P}, \bar{P}, \bar{P}\right\}$ and the left side by Δ_3^T, we obtain the LMI stabilization condition stated in the following theorem.

Theorem 2 *The system (13.17) is α-stabilizable if there exist a real number $\varepsilon > 0$ and $n \times n$ matrices $\bar{P}_1 > 0$, $\bar{P}$, $\bar{Q} = \bar{Q}^T$, $\bar{R} = \bar{R}^T$, and Y such that for all $i \in I^r$ the following LMI is satisfied*

$$\begin{bmatrix} \Phi_i & \begin{bmatrix} e^{\alpha\tau_1}\vartheta \\ \varepsilon e^{\alpha\tau_1}\vartheta \end{bmatrix} & \begin{bmatrix} e^{\alpha\tau_1} D\bar{P} \\ \varepsilon e^{\alpha\tau_1} D\bar{P} \end{bmatrix} \\ * & -\bar{R}/\tau_1 & 0 \\ * & * & -\bar{Q} \end{bmatrix} < 0, \quad \Phi_i = \begin{bmatrix} \Phi_{11} & \Phi_{12} \\ * & \Phi_{22} \end{bmatrix},$$

$$\begin{aligned}
\vartheta &= \left(A_{i\tau_1} - \alpha D\right) \bar{P} + BY, \\
\Phi_{11} &= \left(A_i + \alpha I_n + e^{\alpha\tau_1}\left(A_{i\tau_1} - \alpha D\right)\right) \bar{P} \\
&\quad + \bar{P}^T \left(A_i + \alpha I_n + e^{\alpha\tau_1}\left(A_{i\tau_1} - \alpha D\right)\right)^T + BY + Y^T B^T, \\
\Phi_{12} &= \bar{P}_1^T - \bar{P} + \varepsilon \bar{P}^T \left(A_i + \alpha I_n + e^{\alpha\tau_1}\left(A_{i\tau_1} - \alpha D\right)\right)^T + \varepsilon Y^T B^T, \\
\Phi_{22} &= -\varepsilon\left(\bar{P} + \bar{P}^T\right) + \tau_1 \bar{R} + \bar{Q}.
\end{aligned}$$

Moreover, the feedback gain is given by $K = Y\bar{P}^{-1}$.

In order to apply the above result to the drilling system, we obtain a polytopic representation of the model (13.3), (13.4) and (13.5). The next nonlinear expression to describe the drilling behavior at the ground level is obtained by considering $\mu_{\rm cb} = 0$,

$$\begin{aligned}
&\dot{z}(t) - \Upsilon \dot{z}(t - 2\Gamma) + \left(\Psi + \frac{c_{\rm b}}{I_{\rm B}}\right) z(t) + \Upsilon\left(\Psi - \frac{c_{\rm b}}{I_{\rm B}}\right) z(t - 2\Gamma) \\
&= -c_2 e^{-\frac{\gamma_{\rm b}}{v_{\rm f}} z(t)} \operatorname{sign}(z(t)) + \Upsilon c_2 e^{-\frac{\gamma_{\rm b}}{v_{\rm f}} z(t-2\Gamma)} \operatorname{sign}(z(t - 2\Gamma)) + \Pi u(t - \Gamma),
\end{aligned} \tag{13.19}$$

where $c_2 = \frac{W_{\rm ob} R_{\rm b} \mu_{\rm sb}}{I_{\rm B}}$. We choose the following change of variables:

$$\begin{cases} x_1(t) = z(t), \\ x_2(t) = e^{-\frac{\gamma_{\rm b}}{v_{\rm f}} x_1(t)}, \end{cases} \quad \text{therefore,} \quad \begin{cases} \dot{x}_1(t) = \dot{z}(t), \\ \dot{x}_2(t) = -\frac{\gamma_{\rm b}}{v_{\rm f}} \dot{x}_1(t) x_2(t). \end{cases}$$

System (13.19) can be written as

$$\dot{x}(t) - D\dot{x}(t - 2\Gamma) = A(x)x(t) + A_{2\Gamma}(x)x(t - 2\Gamma) + B_\Gamma u(t - \Gamma), \tag{13.20}$$

where $x = (x_1 \;\; x_2^T)$, and

$$D = \begin{bmatrix} \Upsilon & 0 \\ 0 & 0 \end{bmatrix}, \quad A_{2\Gamma}(x) = \begin{bmatrix} \Upsilon\left(\frac{c_{\rm b}}{I_{\rm B}} - \Psi\right) & c_2 \Upsilon \operatorname{sign}(x_1(t - 2\Gamma)) \\ 0 & 0 \end{bmatrix},$$

$$B_\Gamma = \begin{bmatrix} \Pi \\ 0 \end{bmatrix}, \quad A(x) = \begin{bmatrix} -\left(\Psi + \frac{c_{\rm b}}{I_{\rm B}}\right) & -c_2 \operatorname{sign}(x_1(t)) \\ 0 & -\frac{\gamma_{\rm b}}{v_{\rm f}} \dot{x}_1(t) \end{bmatrix}.$$

Table 13.1 Physical parameters

Symbol	Value	Unit	Symbol	Value	Unit
G	79.3×10^9	N m^{-2}	I	0.095	kg m
L	1172	m	J	1.19×10^{-5}	m^4
R_b	0.155575	m	ν_f	1	∅
W_{ob}	97347	N	I_B	89	kg m^2
c_a	2000	N m s	μ_{cb}	0.5	∅
c_b	0.03	N m s rad^{-1}	c_1	0.3	∅
γ_b	0.9	∅			

Notice that the matrices D, B_Γ are constant, and the entry $c_2\Upsilon \operatorname{sign}(x_1(t-2\Gamma))$ of the matrix $A_{2\Gamma}(x)$ is bounded. If we consider that $\dot{x}_1(t)$ is a bounded variable then, so is the matrix $A(x)$. In this case, we can obtain the polytopic representation:

$$A(x)x(t) + A_{2\Gamma}(x)x(t-2\Gamma) = \sum\nolimits_{i \in I^r} h_i(x_t)\left(A_i x(t) + A_{i2\Gamma}x(t-2\Gamma)\right), \tag{13.21}$$

where A_i, $A_{i2\Gamma}$ have only constant coefficients.

Using the parameters given in Table 13.1, the matrices $A(x)$, $A_{2\Gamma}(x)$, B_Γ and D of the oilwell drilling model (13.20) take the following values,

$$D = \begin{bmatrix} 0.7396 & 0 \\ 0 & 0 \end{bmatrix}, \quad A_{2\Gamma}(x) = \begin{bmatrix} -2.4878 & 100.6802\operatorname{sign}(x_1(t-2\Gamma)) \\ 0 & 0 \end{bmatrix},$$
$$B_\Gamma = \begin{bmatrix} 5.8523 \\ 0 \end{bmatrix}, \quad A(x) = \begin{bmatrix} -3.3645 & -136.1327\operatorname{sign}(x_1(t)) \\ 0 & -0.9\dot{x}_1(t) \end{bmatrix}.$$

Under the assumption that $A(x)$ and $A_{2\Gamma}(x)$ are bounded, we can obtain a polytopic representation in the form (13.21), where $i \in I^r = 2^3 = 8$. For $A(x)$, we have

$$A_i(x) = \begin{bmatrix} -3.3645 & a_{23}^i(x) \\ 0 & a_{33}^i(x) \end{bmatrix},$$
$$-136.1327 = a_{23}^1 \le a_{23}^i(x) \le a_{23}^2 = 0,$$
$$-0.9\mathrm{Acc}_{\max} = a_{33}^1 \le a_{33}^i(x) \le a_{33}^2 = -0.9Dec_{\max},$$

where $Acc_{\max}$ and $Dec_{\max}$ stand for the maximum acceleration and deceleration respectively. For $A_{2\Gamma}(x)$, we have

$$A_{i2\Gamma}(x) = \begin{bmatrix} -2.4878 & a_{2\Gamma 23}^i(x) \\ 0 & 0 \end{bmatrix},$$
$$0 = a_{2\Gamma 23}^1(x) \le a_{2\Gamma 23}^i(x) \le a_{2\Gamma 23}^2(x) = 100.6802.$$

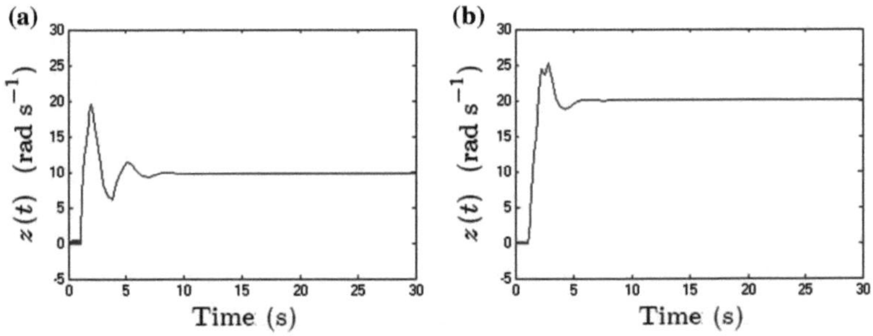

Fig. 13.3 Angular velocity trajectory $z(t)$ of the drilling system (13.19) in closed loop with the control law (13.22) for: **a** $\Omega_0 = 10\,\text{rad s}^{-1}$, **b** $\Omega_0 = 20\,\text{rad s}^{-1}$

Applying the result of Theorem 2 to system (13.11) in closed loop with the control law $u(t) = Kx(t - \Gamma)$, we obtain the feasible result: $K = Y\bar{P}^{-1} = (0.44 \quad -4.25)$. Then, the stabilizing control law for the drilling system (13.19) is given by

$$u(t) = 0.44z(t) - 4.25e^{-\frac{\gamma_b}{\gamma_f}z(t)}. \tag{13.22}$$

The simulation result of Fig. 13.3 shows the expected exponential convergence of the variable $z(t)$ $(x_1(t))$ of the system (13.19) in closed loop with the control law (13.22).

13.4 Conclusion

Two modeling strategies to describe torsional dynamics of a vertical drillstring were presented. The wave equation subject to mixed boundary conditions represents the distributed behavior of the system. Through the D'Alembert method, a nonlinear neutral-type time-delay equation relating certain variables at both extremities of the string was obtained. The input-output model constitutes a reliable representation of the system dynamics simplifying the development of analysis and simulation. Based on both models, the stabilization of the system was investigated; the proposed controllers are aimed at eliminating torsional drillstring vibrations. The control methodology based on a dissipativity analysis of the wave equation model was carried out in a simple and straightforward way, meanwhile the control strategy based on a polytopic representation of the neutral-type time-delay model requires the feasible solution of a set of linear matrix inequalities. Simulation results show a satisfactory performance of the proposed controllers, nevertheless, regarding the response of the closed loop trajectories and the time taken to suppress the stick-slip, we conclude that the controller derived from the multi-model approximation approach is the best solution to suppress torsional drilling oscillations.

References

1. Abolinia, V.E., Myshkis, A.D.: A mixed problem for an almost linear hyperbolic system in the plane. Matematicheskii Sbornik **50**(92), 423–442 (1960)
2. Balanov, A.G., Janson, N.B., McClintock, P.V.E., Wang, C.H.T.: Bifurcation analysis of a neutral delay differential equation modelling the torsional motion of a driven drill-string. Chaos, Solitons Fractals **15**(2), 381–394 (2002)
3. Barton, D.A.W., Krauskopf, B., Wilson, R.E.: Homoclinic bifurcations in a neutral delay model of a transmission line oscillator. Nonlinearity **20**(4), 809–829 (2007)
4. Blakely, J.N., Corron, N.J.: Experimental observation of delay-induced radio frequency chaos in a transmission line oscillator. Chaos **14**(4), 1035–1041 (2004)
5. Boussaada, I., Mounier, H., Niculescu, S.-I., Cela, A.: Analysis of drilling vibrations: a time delay system approach. In: 20th Mediterranean Conference on Control and Automation MED, Barcelona, Spain (2012)
6. Challamel, N.: Rock destruction effect on the stability of a drilling structure. J. Sound Vibr. **233**(2), 235–254 (2000)
7. Dunayevsky, V., Abbassian, F., Judzis, A.: Dynamic stability of drillstrings under fluctuating weight on bit. SPE Drilling Completion **8**(2), 84–92 (1993)
8. Finnie, I., Bailey, J.J.: An experimental study of drill-string vibration. J. Eng. Ind., Trans. ASME **82**(2), 129–135 (1960)
9. Fliess, M., Mounier, H., Rouchon, P., Rudolph, J.: Controllability and motion planning for linear delay systems with an application to a flexible rod. In: Proceedings of the 34th Conference on Decision & Control, TA16 10:40 New Orleans, LA (1995)
10. Fridman, E.: New Lyapunov-Krasovskii functionals for stability of linear retarded and neutral type systems. Syst. Control Lett. **43**(4), 309–319 (2001)
11. Fridman, E., Mondié, S., Saldivar, M.B.: Bounds on the response of a drilling pipe model. Special issue on Time-Delay Systems in IMA J. Math. Control. Inf. **27**(4), 513–526 (2010)
12. Jansen, J.D., van den Steen, L.: Active damping of self-excited torsional vibrations in oil well drillstrings. J. Sound Vibr. **179**(4), 647–668 (1995)
13. Macpherson, J.D., Mason, J.S., Kingman, J.E.E.: Surface measurement and analysis of drill-string vibrations while drilling. SPE/IADC 25777, Society of Petroleum Engineers (1993)
14. Mason, J.S., Sprawls, B.M.: Addressing BHA whirl: the culprit in mobile bay. SPE Drilling Completion **13**(4), 231–236 (1998)
15. Navarro-López, E., Suárez, R.: Practical approach to modelling and controlling stick-slip oscillations in oilwell drillstrings. In: Proceedings of the 2004 IEEE International Conference on Control Applications, pp. 1454–1460 (2004)
16. Pavone, D.R., Desplans, J.P.: Application of high sampling rate downhole measurements for analysis and cure of stick-slip in drilling. In: SPE Annual Technical Conference and Exhibition, SPE 28324, New Orleans, LA, pp. 335–345 (1994)
17. Saldivar, M.B., Mondié, S., Loiseau, J.J., Rasvan, V.: Stick-slip oscillations in oillwell drilstrings: distributed parameter and neutral type retarded model approaches. In: 18th IFAC World Congress Milano, Italy, pp. 284–289 (2011)
18. Sananikone, P.: Method and apparatus for determining the torque applied to a drillstring at the surface. U.S. Patent No. 5,205,163, Schlumberger Technology Co., Houston, TX (1993)
19. Seuret, A.: Commande et observation des systèmes à retard variables, théorie et applications. Thesis PhD, Ecole Centrale de Lille, Univiersité des Sciences et Technologies de Lille, pp. 57–94 (2006)
20. Skaugen, E.: The effects of quasi-random drill bit vibrations upon drillstring dynamic behavior. SPE 16660, Society of Petroleum Engineers (1987)
21. Serrarens, A.F.A., van de Molengraft, M.J.G., Kok, J.J., van den Steeen, L.: $\mathscr{H}_\infty$ control for suppressing stick-slip in oil well drillstrings. IEEE Control Syst. **18**(2), 19–30 (1998)
22. Tucker, R.W., Wang, C.: Torsional vibration control and Cosserat dynamics of a drill-rig assembly. Mecanica **38**(1), 145–161 (2003)

Chapter 14
Flatness-Based Control for a Non-Linear Spatially Distributed Model of a Drilling System

Torsten Knüppel, Frank Woittennek, Islam Boussaada, Hugues Mounier and Silviu-Iulian Niculescu

Abstract The main purpose of this study is the control of both axial and torsional vibrations occurring along a rotary oil well drilling system. The considered model consists of a system of wave equations with non-linear coupled boundary conditions. We propose a flatness-based control approach for suppressing harmful dynamics. Moreover, numerical simulations illustrate the efficiency of the established control laws.

14.1 Introduction

The modelling, analysis and control of rotary drilling vibrations are topics whose economical interest has been renewed by recent oilfields discoveries leading to a growing literature, see for instance [13] investigating the cause of stick-slip vibrations in drilling systems, [7] for the modelling of axial/torsional vibrations occurring along

T. Knüppel (✉) · F. Woittennek
Institute of Control Theory, TU Dresden, 01062 Dresden, Germany
e-mail: torsten.knueppel@tu-dresden.de

F. Woittennek
e-mail: frank.woittennek@tu-dresden.de

H. Mounier · S.-I. Niculescu
Laboratoire des Signaux et Systèmes, CNRS, Supélec, Université Paris Sud,
3 rue Joliot-Curie, 91192 Gif-sur-Yvette Cedex, France
e-mail: hugues.mounier@lss.supelec.fr

S.-I. Niculescu ·
e-mail: silviu.niculescu@lss.supelec.fr

I. Boussaada
IPSA & Laboratoire des Signaux et Systèmes, 3 rue Joliot-Curie,
91192 Gif-sur-Yvette Cedex, France
e-mail: islam.boussaada@lss.supelec.fr

A. Seuret et al. (eds.), *Low-Complexity Controllers for Time-Delay Systems*,
Advances in Delays and Dynamics 2, DOI: 10.1007/978-3-319-05576-3_14,

a rotary drilling system, [6] for the control of drilling vibrations using a Lyapunov functional approach, [12] for the control of the drilling torsional vibrations modeled by a finite dimensional dynamical system, and [17] for a control approach avoiding the stick-slip oscillations occurring in the drilling process.

Roughly speaking, a rotary drilling structure consists essentially of a rig, a drill string, and a bit. The essential components of the drill string are the bottom hole assembly, composed mainly of heavy steel tubes to provide a large downward force on the bit, and a set of drill pipes made of thinner tubes. The drill string is in particular subject to two main types of vibrations, each of them can at the least cause a premature wear of the various components. Torsional vibrations are responsible of the so-called stick slip phenomenon which is essentially the cause of premature breakage of the drill pipes. Traction/compression vibrations, or shortly axial vibrations, mainly associated with the bit bouncing phenomenon may cause premature wear of the bit.

These two types of vibrations are known to be coupled and present several non-linear phenomena. This combination makes the derivation of a model mathematically challenging and is the reason why the full system has been rarely considered so far. For the sake of reducing the complexity of the problem some works suggest that the axial vibrations are neglected with respect to torsional vibrations in order to avoid considering coupling dynamics (cf. [1]). Another approach is to neglect the infinite-dimensional aspects of those vibrations by using a lumped parameter model consisting of ordinary differential equations (cf. [11, 12]). Finally, the non-linearity of the boundary conditions which comes essentially from the friction profile at the bit is approximated by a piecewise linear function (cf. [7]). Unfortunately, the adoption of such simplifications impoverishes the recovered dynamics.

Throughout this contribution we consider a model established in [2], which takes into account all the aforementioned aspects. More precisely, the propagation of each type of vibration is assumed to be governed by a wave equation with non-linear boundary conditions. Moreover, it is emphasized that the coupling terms in the boundary conditions are induced by the interface bit/rock friction.

In the following, the design of control laws for both axial and torsional vibrations is discussed. To this end, a flatness-based feed-back controller is proposed that guarantees a regular drilling process by suppressing the undesired dynamics. The originality of the present paper does not lie only in the used control strategy itself, since a flatness-based feedback controller is derived in [14] and [10] for stabilizing the torsional vibrations governed by a wave equation with simpler boundary conditions, but also in applying it to a relatively complete model.

14.2 Model of a Drill String

For the following considerations a model proposed in [2] is used to describe the axial and torsional vibrations, i.e. the behaviour of the torsion angle ϕ and the axial displacement U. These two system variables are spatially and timely dependent, where the spatial variable s is chosen such that $s = 0$ denotes the top of the drill string and $s = \ell$ its bottom (see Fig. 14.1).

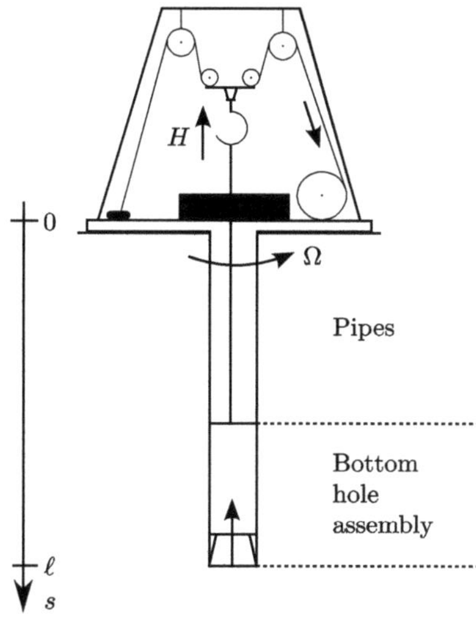

Fig. 14.1 Sketch of the drill string

The model comprises the two wave equations

$$\frac{\partial^2 U}{\partial t^2}(s,t) = c^2 \frac{\partial^2 U}{\partial s^2}(s,t), \tag{14.1}$$

and

$$\frac{\partial^2 \phi}{\partial t^2}(s,t) = \tilde{c}^2 \frac{\partial^2 \phi}{\partial s^2}(s,t), \tag{14.2}$$

with boundary conditions

$$E\Gamma \frac{\partial U}{\partial s}(0,t) = \alpha \frac{\partial U}{\partial t}(0,t) - H(t), \tag{14.3a}$$

$$M \frac{\partial^2 U}{\partial t^2}(\ell,t) = -E\Gamma \frac{\partial U}{\partial s}(\ell,t) - pF\left(\frac{\partial U}{\partial t}(\ell,t)\right), \tag{14.3b}$$

and

$$G\Sigma \frac{\partial \phi}{\partial s}(0,t) = \beta \frac{\partial \phi}{\partial t}(0,t) - \Omega(t), \tag{14.4a}$$

$$J \frac{\partial^2 \phi}{\partial t^2}(\ell,t) = -G\Sigma \frac{\partial \phi}{\partial s}(\ell,t) - \tilde{p}F\left(\frac{\partial U}{\partial t}(\ell,t)\right). \tag{14.4b}$$

The speeds of propagation c and $\tilde{c}$ can be computed from material parameters, namely Young's modulus E, the shear modulus G and the density ρ, by means of

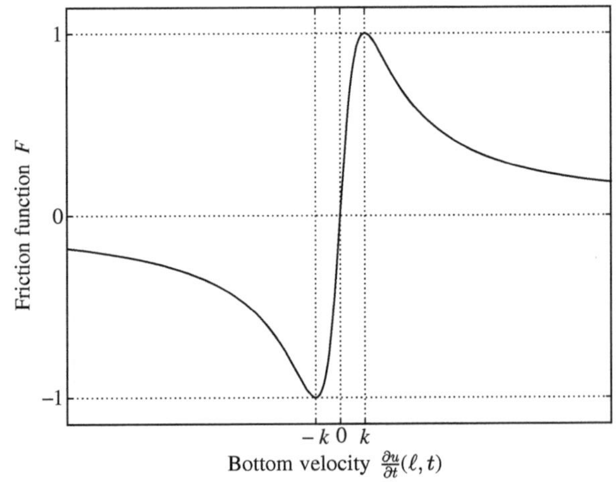

Fig. 14.2 Graph of friction function F

$$c = \sqrt{\frac{E}{\rho}} \quad \text{and} \quad \tilde{c} = \sqrt{\frac{G}{\rho}}. \tag{14.5}$$

Apart from these physical properties the model contains geometrical parameters of the drill string, that are assumed to be spatially and timely constant. These comprise the drill string's cross-section Γ and its second moment of area Σ, as well as the mass M and the moment of inertia J of the drill bit. The two parameters p and $\tilde{p}$ together with the function F appearing in the boundary conditions at the bottom account for the friction resulting from the interaction between the drill bit and the rock. The function F is assumed to be of the form

$$F(x) = \frac{2kx}{x^2 + k^2},$$

where k is a constant, positive parameter (see Fig. 14.2). The system can be controlled by the boundary force H and the boundary torque Ω .

14.3 Flatness of the Drill String Model

In this section it will be shown that the previously introduced model can be considered flat. Even though flatness is a system property that has been originally defined for finite-dimensional models (cf. [5]), its basic idea can be extended to the infinite-dimensional case. This idea consists in the possibility to parametrize all system variables by means of a so-called flat output, i.e. once a trajectory for the flat output has been prescribed the trajectories of all system variables can be computed from it. The main difference between the finite and the infinite-dimensional case lies in the character of the relation between the flat output and the system variables: While in the first case this relation involves only finite-order derivatives of the flat output, it might comprise derivatives of arbitrary order or delays and predictions in the latter case depending on the type of the underlying partial differential equation.

In the following a flatness-based parametrization of the drill string model will be derived. First of all, this will be done for the subsystem describing the axial vibrations, i.e. Eq. (14.1) with boundary conditions (14.3), as it is independent of the torsional vibrations. Because of the structural similarity of both subsystems, the considerations made in this first step, will greatly simplify the analysis of the entire system.

14.3.1 Flatness of the Axial Subsystem

In similar examples the boundary value of the distributed system variable at the unactuated boundary, i.e. $U(\ell, \cdot)$ in the present case, served as flat output. It will be shown that this is the case here, too.

To this end, a parametrization of all system variables appearing in the axial subsystem, i.e. the distributed variable U and the control input H, by the boundary values of U and $\frac{\partial U}{\partial s}$ at $s = \ell$ is derived. Afterwards it will be proven that these values and, thus, the whole solution can be computed from the flat output $y_U = U(\ell, \cdot)$.

The parametrization of the axial displacement U is obtained by solving the partial differential equation (14.1). This can be done for example by means of the classical d'Alembert solution, the method of characteristics or operational calculus and yields

$$U(s,t) = \frac{1}{2}\left(U\left(\ell, t + \frac{\ell - s}{c}\right) + U\left(\ell, t - \frac{\ell - s}{c}\right) - c\int_{t-\frac{\ell-s}{c}}^{t+\frac{\ell-s}{c}} \frac{\partial U}{\partial s}(\ell, \tau)\, d\tau\right). \tag{14.6}$$

It becomes obvious that the solution depends on predicted and delayed values of the involved boundary trajectories. Substituting this solution into the boundary condition (14.3a) leads to a parametrization

$$\begin{aligned} H(t) =& h_+\left(\frac{\partial U}{\partial t}\left(\ell, t + \frac{\ell}{c}\right) - c\frac{\partial U}{\partial s}\left(\ell, t + \frac{\ell}{c}\right)\right) \\ &+ h_-\left(\frac{\partial U}{\partial t}\left(\ell, t - \frac{\ell}{c}\right) + c\frac{\partial U}{\partial s}\left(\ell, t - \frac{\ell}{c}\right)\right), \end{aligned} \tag{14.7}$$

of the force H, where

$$h_+ = \frac{1}{2}\left(\alpha + \frac{E\Gamma}{c}\right) \quad \text{and} \quad h_- = \frac{1}{2}\left(\alpha - \frac{E\Gamma}{c}\right),$$

are constants. Thus, it has been shown that all involved system variables can be parametrized by the boundary values $U(\ell, \cdot)$ and $\frac{\partial U}{\partial s}(\ell, \cdot)$. It remains to show that these boundary values can be computed from the flat output $y_U = U(\ell, \cdot)$. As this is obvious for $U(\ell, \cdot)$, it remains to establish a relation between the gradient $\frac{\partial U}{\partial s}$ and y_U. This can be obtained from the boundary condition (14.3b):

$$\frac{\partial U}{\partial s}(\ell, t) = -\frac{1}{E\Gamma}\left(F\left(\frac{dy_U}{dt}(t)\right) + M\frac{d^2 y_U}{dt^2}(t)\right). \tag{14.8}$$

14.3.2 Flatness of the Entire System

It can be observed from the model equations that the torsional and the axial subsystem are structurally similar except for the boundary conditions at the bottom. Thus, the results from the previous section suggest that the boundary value $\phi(\ell, \cdot)$ might be a flat output of the torsional system, if the axial displacement U is considered known. This would amount to a flat output of the entire system comprising the boundary values of both distributed system variables, i.e. $y = (y_U, y_\phi) = (U(\ell, \cdot), \phi(\ell, \cdot))$. It will be shown in the following that y is indeed a flat output of the entire system. As it has already been shown in the previous section that the axial displacement U and the force H can be parametrized by the first component y_U of y, it suffices to consider the angle ϕ and the torque Ω.

The flatness-based parametrization of the torsional model is obtained along the same line as the one of the axial subsystem from a representation of the solution of the governing partial differential equation (14.2) in terms of the boundary values at $s = \ell$:

$$\phi(s, t) = \phi\left(\ell, t + \frac{\ell - s}{\tilde{c}}\right) + \phi\left(\ell, t - \frac{\ell - s}{\tilde{c}}\right) - \tilde{c}\int_{t-\frac{\ell-s}{\tilde{c}}}^{t+\frac{\ell-s}{\tilde{c}}} \frac{\partial \phi}{\partial s}(\ell, \tau)\, d\tau. \tag{14.9}$$

Substituting this parametrization into the boundary condition (14.4a) yields an expression of the torque Ω in terms of the boundary values and the two constants

$$\omega_+ = \frac{1}{2}\left(\beta + \frac{G\Sigma}{\tilde{c}}\right) \quad \text{and} \quad \omega_- = \frac{1}{2}\left(\beta - \frac{G\Sigma}{\tilde{c}}\right),$$

that reads

$$\begin{aligned} \Omega(t) = &\,\omega_+\left(\frac{\partial \phi}{\partial t}\left(\ell, t + \frac{\ell}{\tilde{c}}\right) - \tilde{c}\frac{\partial \phi}{\partial s}\left(\ell, t + \frac{\ell}{\tilde{c}}\right)\right) \\ &+ \omega_-\left(\frac{\partial \phi}{\partial t}\left(\ell, t - \frac{\ell}{\tilde{c}}\right) + \tilde{c}\frac{\partial \phi}{\partial s}\left(\ell, t - \frac{\ell}{\tilde{c}}\right)\right). \end{aligned} \tag{14.10}$$

Similar to the previous section, it has to be shown that the boundary values can be computed from the flat output y. To this end, it suffices to obtain a relation between the gradient $\frac{\partial \phi}{\partial s}$ and y. This can be derived from the boundary condition (14.4b):

$$\frac{\partial \phi}{\partial s}(\ell, t) = -\frac{1}{G\Sigma}\left(\tilde{F}\left(\frac{dy_U}{dt}(t)\right) + J\frac{d^2 y_\phi}{dt^2}(t)\right). \tag{14.11}$$

It has been shown that the model proposed in Sect. 14.2 is flat and the boundary values of the distributed system variables, i.e. $y = (U(\ell, \cdot), \phi(\ell, \cdot))$, compose a flat output.

14.4 Flatness-Based Control Design

In this section the derivation of both open- and closed-loop control laws will be discussed for the considered drilling system. As the open-loop case is easily solved using the flatness-based parametrizations obtained in the previous section, the focus is on the feedback control design.

14.4.1 Open-Loop Control

The flatness-based parametrization obtained in the previous section can be directly used for the design of a feed-forward controller. To this end, it suffices to prescribe an appropriate trajectory for the flat output y and compute the required control inputs from it. Inspecting Eqs. (14.6)–(14.11) reveals that the highest derivative of y involved is of order 2. Thus, in order to obtain continuous trajectories of the system variables the flat output has to be twice continuously differentiable. Details have been presented in numerous publications and are, therefore, omitted for the sake of brevity (cf. [10, 16]).

14.4.2 Feedback Control

The main idea in designing a flatness-based feed-back controller is to compute the control inputs such that the error between the desired and the actual trajectory of the flat output satisfies an asymptotically stable dynamic. The question that remains is how to define an appropriate error taking into account the different delays and predictions appearing in the flatness-based parametrization. In the following the tracking error is defined for the values with the largest predictions (cf. [9] for a similar approach concerning delay systems). By inspecting the flatness-based parametrization it becomes obvious that the components y_U and y_ϕ of the flat output appear with different predictions, namely $T = \frac{\ell}{c}$ and $\tilde{T} = \frac{\ell}{\tilde{c}}$. As the shear module G is assumed to be less than Young's modulus E and, thus, $\tilde{c}$ is less than c according to (14.5), it follows that T is less than $\tilde{T}$. Thus, the appropriate error reads

$$e(t) = y(t + \tilde{T}) - y_d(t + \tilde{T}),$$

where y_d is the desired trajectory of the flat output. For this error a stable dynamic of the form

$$\frac{d^2 e}{dt^2} + A_1 \frac{de}{dt} + A_0 e = 0,$$

is prescribed. To ensure the stability of this dynamic the matrices A_0 and A_1 have to be chosen such that the eigenvalues of the matrix

$$\begin{bmatrix} 0_{2\times 2} & I_{2\times 2} \\ -A_0 & -A_1 \end{bmatrix},$$

lie in the open left half plane, where $0_{2\times 2}$ and $I_{2\times 2}$ denote the zero and the identity matrix respectively. Once these matrices have been chosen the error dynamic can be used to compute the predicted value of the flat output's second derivative:

$$\frac{d^2 y}{dt^2}(t + \tilde{T}) = \frac{d^2 y_d}{dt^2}(t + \tilde{T}) - A_1 \frac{de}{dt}(t) - A_0 e(t). \tag{14.12}$$

It becomes obvious from this equation that predicted values of y and its first derivative are required for the computation of the second derivative. In the next section a prediction algorithm will be presented that makes these values available. Thus, they are assumed to be known for the remainder of this section.

The trajectories of the controls required for the realization of the stable error dynamics can be obtained by substituting the predicted values obtained from the stable error dynamic (14.12) and the prediction algorithm into the respective Eqs. (14.7) and (14.10).

An inspection of these two equations reveals that they involve delayed values. The most obvious method for their computation is to obtain them from interpolating saved measurements. However, it turned out to be advantageous to eliminate the delayed values by means of additional measurements at the top of the drill string. To this end, one obtains from Eqs. (14.6) and (14.9):

$$\frac{\partial U}{\partial t}\left(\ell, t - \frac{\ell}{c}\right) + c\frac{\partial U}{\partial s}\left(\ell, t - \frac{\ell}{c}\right) = \frac{\partial U}{\partial t}(0, t) + c\frac{\partial U}{\partial s}(0, t),$$

and

$$\frac{\partial \phi}{\partial t}\left(\ell, t - \frac{\ell}{\tilde{c}}\right) + \tilde{c}\frac{\partial \phi}{\partial s}\left(\ell, t - \frac{\ell}{\tilde{c}}\right) = \frac{\partial \phi}{\partial t}(0, t) + \tilde{c}\frac{\partial \phi}{\partial s}(0, t).$$

Remark 1 Following the considerations in [19] and [20] the control law obtained above can be interpreted as a state feed-back. To make this plausible, only the axial subsystem is considered first: Inspecting Eqs. (14.6) and (14.7) reveals that there is a transformation between the physical state, that comprises the profiles of $U(\cdot, t)$ and $\frac{\partial U}{\partial s}(\cdot, t)$ as well as the flat output values $y_U(t)$ and $\frac{dy_U}{dt}(t)$, and the restriction of the flat output's trajectory to the interval $[t-T, t+T]$. Thus, the restriction of y_U to the interval $[t-T, t+T]$ can be considered a state of the axial subsystem. If now the whole subsystem is considered it becomes apparent that there is a transformation between the complete physical state, i.e. the above physical state of the axial subsystem together with the equivalent values of the torsional subsystem, and certain restrictions of the components of the flat output, namely $[t-T, t+T] \ni \tau \mapsto y_U(\tau)$ and $[t-\tilde{T}, t+\tilde{T}] \ni \tau \mapsto y_\phi(\tau)$. However, this transformation requires values of y_U outside the given restriction which becomes apparent from Eqs. (14.10) and (14.11). Therefore, predicted and delayed values of the control input Ω are required in the course of this transformation, that is, thus, not a classical one. Consequently the state comprising the above restrictions of the flat output's trajectory is not a classical state.

These considerations suggest that the above derived control law on the basis of the stable dynamic (14.12) is not a static feed-back. It can be categorized as a dynamic feed-back that was obtained by extending the component of the state that corresponds to the axial subsystem. Thus, by choosing the error $e(t) = y(t+\tilde{T}) - y_d(t+\tilde{T})$ instead of

$$e(t) = \begin{pmatrix} y_U(t+T) - y_{U,d}(t+T) \\ y_\phi(t+\tilde{T}) - y_{\phi,d}(t+\tilde{T}), \end{pmatrix}$$

one obtained the new input $v(t) = \frac{d^2y}{dt^2}(t+\tilde{T})$ instead of

$$v(t) = \begin{pmatrix} \frac{d^2y_U}{dt^2}(t+T) \\ \frac{d^2y_\phi}{dt^2}(t+\tilde{T}) \end{pmatrix}.$$

Alternatively, one could use a quasi-static feed-back that required the prediction of the control law. These generalizations of concepts discussed in [4] and [15] for finite-dimensional systems will be subject of further research.

14.5 Prediction of Boundary Values

One way to predict the values required for evaluating the stable error dynamics (14.12) is to compute them from the system state, i.e. the profiles of the system variables $s \mapsto (U(s, \cdot), \phi(s, \cdot))$ and their respective time derivatives. These can be obtained from distributed measurements or by using an observer (cf. [20]). A simpler approach is based on the observation that at time t the values of the second derivative of y are known on the interval $[t, t+\tilde{T})$ from previous steps. Thus, the flat output y and its first derivative can be obtained by (numerically) integrating these known values, i.e. by solving

Table 14.1 Physical parameters

Symbol	Value	Unit
E	200	GPa
ρ	8000	kg/m^3
Γ	35	cm^2
α	200.025	kg s^{-1}
M	40000	kg
p	3500	Ø
k	0.18	Ø
G	80	GPa
ℓ	3500	m
Σ	19	cm^4
β	2000	N ms
J	144	kg m^2
$\tilde{p}$	210	Ø

$$\frac{dy}{dt}(t+\tau) = \frac{dy}{dt}(t) + \int_0^{\tau} \frac{d^2y}{dt^2}(t+\xi)\,d\xi,$$
$$y(t+\tau) = y(t) + \int_0^{\tau} \frac{dy}{dt}(t+\xi)\,d\xi,$$

where $\tau \in [0, \tilde{T}]$. It becomes obvious that this prediction algorithm requires measurements of the flat output and its first derivative at time t. The main drawback of this approach is its close dependence on the model, because it is assumed that the predicted values of the second derivatives are indeed realized. Thus, the prediction algorithm does not take into account errors in these values that occur due to disturbances or model uncertainties. This suggests that the initially sketched approach involving an observer or distributed measurements would yield more precise results.

14.6 Numerical Simulation

The system equations were simulated using a semi-discretisation in conjunction with a Python/SciPy implementation of a numerical integration method called VODE (cf. [3, 8, 18]). To this end, the derivatives w.r.t. the spatial coordinate s were replaced by finite differences. Thus, one obtains a system of ordinary differential equations w.r.t. the time, that is then solved with above mentioned numerical integration method. The numerical values of the physical parameters that are used in the following are given in Table 14.1.

The control problem considered for the simulation study is to steer the system from initial rest to a regime that is characterized by constant torsional and axial velocities. To be precise, it should be at rest until time $t = t_i$ and the transition is supposed to be finished at time $t = t_f$. From the flatness-based parametrization of the solution it can

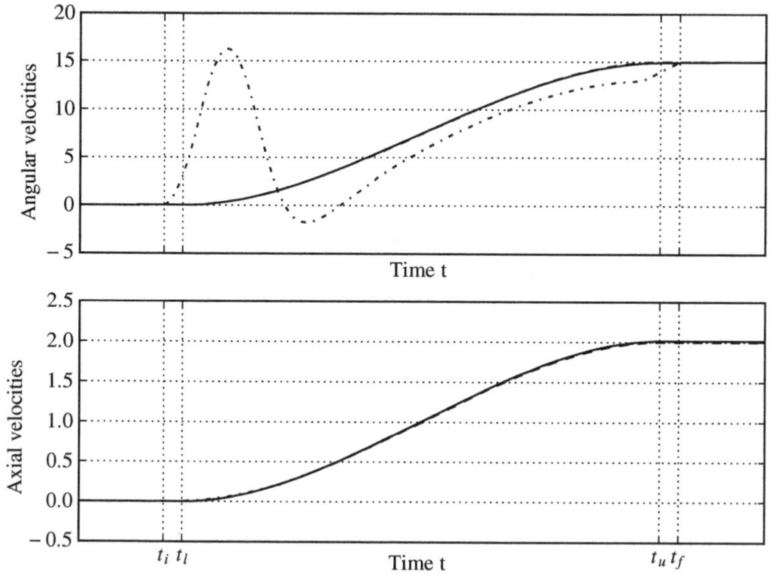

Fig. 14.3 Trajectories of the axial and angular velocity using the feed-forward control approach—velocity at top (*dot-dashed*), velocity at bottom (*solid*), desired trajectory of bottom velocity (*dashed*)

be concluded that this amounts to a desired trajectory of the flat output $t \mapsto y_d(t)$ that is twice continuous differentiable and fulfils the following conditions:

$$\begin{aligned} y_d(t) &= 0, \quad t \leq t_l, \\ \frac{dy_d}{dt}(t) &= v_d, \quad t \geq t_u, \end{aligned} \tag{14.13}$$

where $t_l = t_i + \tilde{T}$ and $t_u = t_f - \tilde{T}$. Such a trajectory is for example given by

$$y_d(t) = \begin{cases} 0, & t \leq t_l, \\ \phi\left(\frac{t-t_l}{t_u-t_l}\right), & t_l < t < t_u, \\ v_d\left(t - \frac{t_l+t_u}{2}\right), & t \geq t_u, \end{cases}$$

with the polynomial

$$\phi(\tau) = v_d(t_u - t_l)\left(1 - \frac{\tau}{2}\right)\tau^3.$$

Figures 14.3 and 14.4 depict the simulation results of the feed-forward controller. It becomes obvious that apart from small deviations, that are due to numerical errors, the desired trajectory of the flat output is followed. Furthermore, the effect of the delays and predictions in the system become apparent in the trajectories of the velocities at

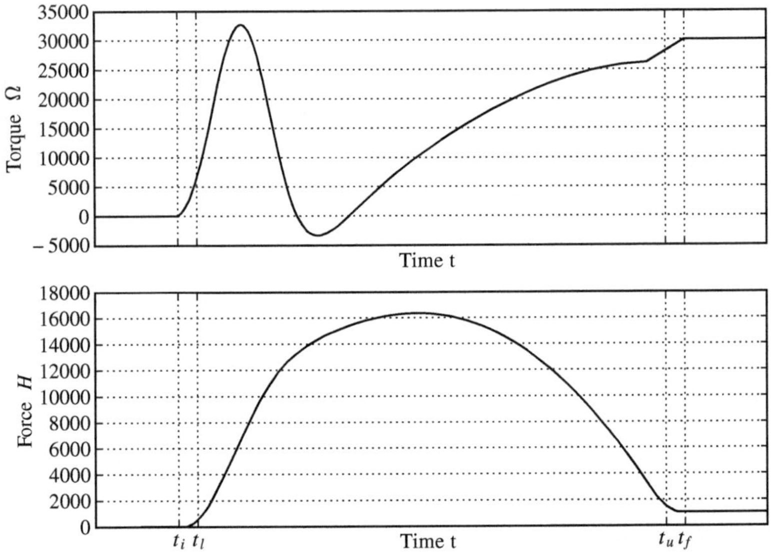

Fig. 14.4 Trajectories of the control inputs using the feed-forward control approach

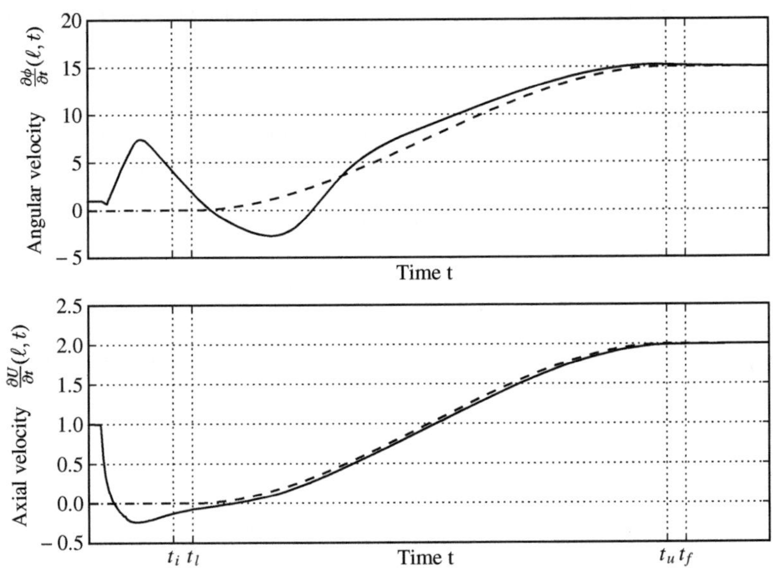

Fig. 14.5 Trajectories of the axial and angular velocity using the feed-back control approach—simulated trajectories (*solid*) and desired trajectories (*dashed*)

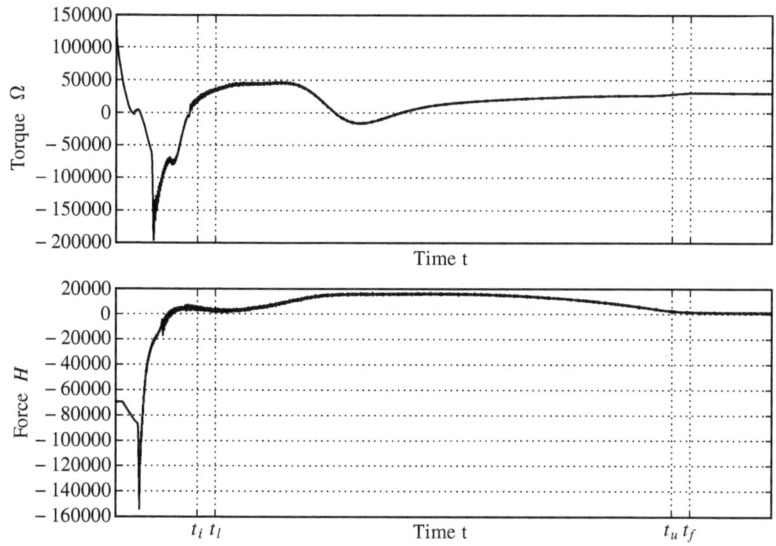

Fig. 14.6 Trajectories of the control inputs using the feed-back control approach

the top of the drill string and the control inputs: These velocities show non-stationary behaviour outside of the interval $[t_l, t_u]$, even though the flat output is held constant there.

The behaviour of the feed-back controller is depicted in Figs. 14.5 and 14.6. The same desired trajectories as in the feed-forward case were used, but the drill string was assumed to be not at rest initially. It becomes apparent that the controller is able to force the system to the desired trajectory.

14.7 Conclusion and Outlook

The nature of the drilling process makes the use of robust control laws inevitable, as there are great uncertainties in the boundary conditions. The presented approach certainly does not match this requirement, because of its close dependence on the used friction model. However, it can be considered a first step towards a realistic control law that yields valuable insights into the system dynamics, e.g. the magnitude of delays and predictions or the coupling mechanisms. Using this knowledge the further research will be directed towards the investigation of the effects of model uncertainties and their compensation. At the same time more elaborate models will be discussed, especially with regard to an appropriate description of the friction effects.

Acknowledgments The scientific exchange was enabled thanks to the financial support of the German Academic Exchange Service (DAAD grant no. 50749065) within the PROCOPE program. The second author is also supported by the DFG (grant no. Wo 1601/1).

References

1. Balanov, A.G., Janson, N.B., McClintock, P.V.E., Wang, C.H.T.: Bifurcation analysis of a neutral delay differential equation modelling the torsional motion of a driven drill-string. Chaos, Solitons Fractals **15**(2), 381–394 (2002)
2. Boussaada, I., Mounier, H., Niculescu, S.-I., Cela, A.: Analysis of drilling vibrations: a time-delay system approach. In: 20th Mediterranean Conference on Control Automation (MED), pp. 610–614 (2012)
3. Brown, P., Byrne, G., Hindmarsh, A.: Vode: a variable-coefficient ode solver. SIAM J. Sci. Stat. Comput. **10**(5), 1038–1051 (1989)
4. Delaleau, E., Rudolph, J.: Control of flat systems by quasi-static feedback of generalized states. Int. J. Control **71**(5), 745–765 (1998)
5. Fliess, M., Lévine, J., Martin, Ph., Rouchon, P.: Flatness and defect of non-linear systems: introductory theory and examples. Int. J. Control **61**(6), 1327–1361 (1995)
6. Fridman, E., Mondié, S., Saldivar, B.: Bounds on the response of a drilling pipe model. IMA J. Math. Control Inf. **27**(4), 513–526 (2010)
7. Germay, C., Van de Wouw, N., Nijmeijer, H., Sepulchre, R.: Nonlinear drillstring dynamics analysis. SIAM J. Appl. Dyn. Syst. **8**(2), 527–553 (2009)
8. Jones, E., Oliphant, T., Peterson, P., et al.: SciPy: open source scientific tools for Python (2001)
9. Mounier, H., Rudolph, J.: Flatness based control of nonlinear delay systems: a chemical reactor example. Int. J. Control **71**(5), 871–890 (1998)
10. Mounier, H., Rudolph, J., Petitot, M., Fliess, M.: A flexible rod as a linear delay system. In: Proceedings of 3rd European Control Conference, pp. 3676–3681 (1995)
11. Navarro-López, E.M.: An alternative characterization of bit-sticking phenomena in a multi-degree-of-freedom controlled drillstring. Nonlinear Anal. Real World Appl. **10**(5), 3162–3174 (2009)
12. Navarro-López, E.M., Cortes, D.: Sliding-mode control of a multi-dof oilwell drillstring with stick-slip oscillations. In: American Control Conference, 2007. ACC'07, pp. 3837–3842 (2007)
13. Richard, T., Germay, C., Detournay, E.: A simplified model to explore the root cause of stick-slip vibrations in drilling systems with drag bits. J. Sound Vibr. **305**(3), 432–456 (2007)
14. Rouchon, P.: Flatness and stick-slip stabilization. Technical Report, Mines-ParisTech **492**, 1–9 (1998)
15. Rudolph, J., Delaleau, E.: Some examples and remarks on quasi-static feedback of generalized states. Automatica **34**(8), 993–999 (1998)
16. Rudolph, J., Winkler, J., Woittennek, F.: Flatness based control of distributed parameter systems: examples and computer exercises from various technological domains. Berichte aus der Steuerungs- und Regelungstechnik. Shaker Verlag, Aachen (2003)
17. Saldivar, B., Mondié, S., Loiseau, J.-J., Rasvan, V.: Stick-slip oscillations in oilwell drilstrings: distributed parameter and neutral type retarded model approaches. In: Proceedings of the 18th IFAC World Congress, pp. 284–289 (2011)
18. Schiesser, W.E.: The Numerical Method of Lines: Integration of Partial Differential Equations. Academic Press, San Diego (1991)
19. Woittennek, F.: On flatness and controllability of simple hyperbolic distributed parameter systems. In: Proceedings of 18th IFAC World Congress, pp. 14452–14457, Milano, Italy (2011)
20. Woittennek, F., Rudolph, J.: Controller canonical forms and flatness based state feedback for 1D hyperbolic systems. In: Preprints 7th Vienna International Conference on Mathematical Modeling (MathMod) (2012)

Chapter 15
Smith Predictor-Based Control with Distance Feedback for Haptic Systems Under Distributed Time-Delays

Bogdan Liacu, Irinel-Constantin Morărescu, Silviu-Iulian Niculescu, Claude Andriot, Didier Dumur, Frédéric Colledani and Patrick Boucher

Abstract This chapter proposes a Smith predictor-based control with distance feedback until a possible collision for network based haptic systems. The main idea is to use a predictor just on the haptic side in order to compensate the viscosity effect and to provide an accurate feeling in case of contacts. The critical problem is that in haptics there are two situations: free and restricted motion, which implies a model changing on the predictor side. For solving this problem, a new approach is presented by using the available information on the distance from the virtual reality simulator and introducing it in the predictor in order to maintain the similitude between the

B. Liacu · S.-I. Niculescu (✉)
Laboratoire des Signaux et Systèmes (L2S), CNRS-SUPELEC, 3 rue Joliot Curie, Gif-sur-Yvette Cedex 91192, France
e-mail: silviu.niculescu@lss.supelec.fr

B. Liacu
e-mail: bogdan.liacu@gmail.com

I.-C. Morărescu
CRAN, UMR 7039 and CNRS, CRAN, UMR 7039, Université de Lorraine, 2 Avenue de la Forêt de Haye, Vandœuvre-lès-Nancy, France
e-mail: constantin.morarescu@univ-lorraine.fr

C. Andriot · F. Colledani
CEA, LIST, Interactive Robotics Laboratory, Fontenay aux Roses 92265, France
e-mail: claude.andriot@cea.fr

F. Colledani
e-mail: frederic.colledani@cea.fr

D. Dumur · P. Boucher
Control Department, SUPELEC E3S, 3 rue Joliot Curie, Gif sur Yvette Cedex 91192, France
e-mail: didier.dumur@supelec.fr

P. Boucher
e-mail: patrick.boucher@supelec.fr

A. Seuret et al. (eds.), *Low-Complexity Controllers for Time-Delay Systems*, Advances in Delays and Dynamics 2, DOI: 10.1007/978-3-319-05576-3_15,

Fig. 15.1 Examples of virtual environment applications. **a** Virtual prototyping, **b** virtual assistance/supervision, **c** virtual assembly

real and the predicted model. The three most common delay distributions (uniform, gamma with gap and normal) are analyzed from the stability point of view and are tested on a 3-dof real-time experimental platform.

15.1 Introduction

In the last decades, virtual environments have become very popular and are used in many domains, like prototyping (Fig. 15.1a example of prototyping using haptic interfaces and virtual environment [1]), training for different devices and assistance in completing difficult tasks (Fig. 15.1b virtual environment used for task assistance/supervision [2]), virtual assembling (Fig. 15.1c—example of virtual assembly [3]). Haptic systems represent an extension of the teleoperation systems. A teleoperation system is composed by two robots, geographically different disposed, working on the principle of master/slave. Generally, the human operator imposes a force and/or a position to the master robot which will transmit the command to the slave robot. Depending on the situations encountered by the slave robot, the master should also be able to *reproduced* them via the feedback force (a complete survey on teleoperation systems can be found in [4]. A haptic system has only the master robot, which is usually called haptic interface, and the second robot is replaced by a virtual robot/object which is working in a virtual environment. The principle is the same, the user should *feel* the contacts encountered by the virtual object/robot via the feedback force.

A major control problem of such systems is the presence of time-delays (see, for instance [5, 6] and the references therein for some references on such topics). The presence of time-delays may induce instabilities and complex behaviors. Systems with distributed delays are present in many scientific fields such as economy [7], population dynamics [8], traffic control [9], biological systems [10], network/Internet-based systems [11], and many others. An useful approach for dealing with time-varying delay problems is the distribution's *modeling*. Roughly speaking, the delay distribution aims to describe the delay variation in some stochastic manner and to capture its average behavior. The distributions that are mostly used in the literature to describe the delay are the uniform, the normal and the gamma-distribution.

In haptics, there are two time-delay sources: the communication channel and the processing time for the virtual reality environment. More precisely, in free motion the delay effect can be felt by the viscosity phenomenon (high force feedback felt at the haptic interface end), in the case of a hard contact with the environment, the impact effect will not be stiff, or the most unwanted situation is to loose the system stability due to the delays. Therefore, the delays have to be taken into account and included in the control laws.

In this chapter, a Smith predictor based control [12] with distance feedback until a possible collision is proposed. The stability analysis method follows the ideas presented in [13, 14]. Generally, the Smith predictor is working correctly (for haptic systems) when the delays are fixed and known and also the model used in the predictor is close to the real model. The central idea is to use a predictor just on the haptic side in order to compensate the viscosity effect and to provide an accurate feeling in case of contacts.

For solving such a problem, a new approach is presented, by using the available information on the distance from the virtual reality simulator and introducing it in the predictor in order to maintain the similitude between the real and the predicted model. This work represents an extended version of the conference proceedings paper [15].

15.2 Theoretical Background

For an ideal haptic system, the following conditions have to be satisfied: (i) position tracking error as small as possible between the haptic interface and the virtual object, (ii) high degree of transparency, i.e. in free motion, the force feedback felt at the haptic interface end must be as small as possible and in case of hard contact, a stiff response is desired. The starting point is the *classical* dynamic (nonlinear) equations of motion for two similar robots in the framework of haptic systems:

$$M_h(x_h)\ddot{x}_h(t) + C_h(x_h, \dot{x}_h)\dot{x}_h = -F_h(t) + F_{op}(t), \tag{15.1}$$

$$M_v(x_v)\ddot{x}_v(t) + C_v(x_v, \dot{x}_v)\dot{x}_v = F_v(t) - F_e(t), \tag{15.2}$$

where x_h, x_v are the haptic interface/virtual object position, F_{op}, F_e are the human/environmental forces, F_h, F_v are the force control signals, M_h, M_v are the symmetric and positive-definite inertia matrices, and C_h, C_v are the Coriolis matrices of the haptic interface and virtual object systems, respectively. The main idea is to use two similar PD controllers, one to control the haptic interface and another one for the virtual object. The controller equations are there given as follows:

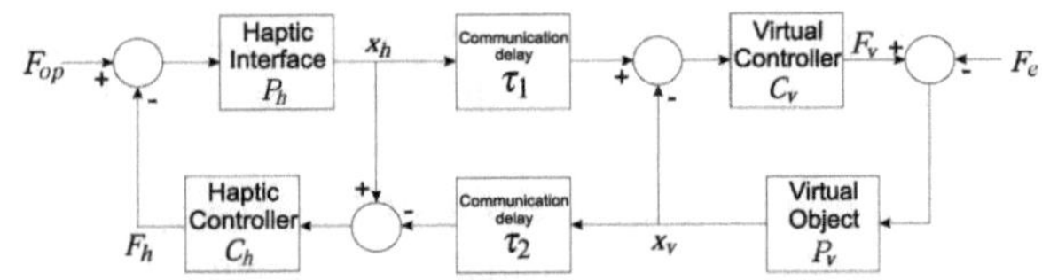

Fig. 15.2 Control scheme of the haptic system

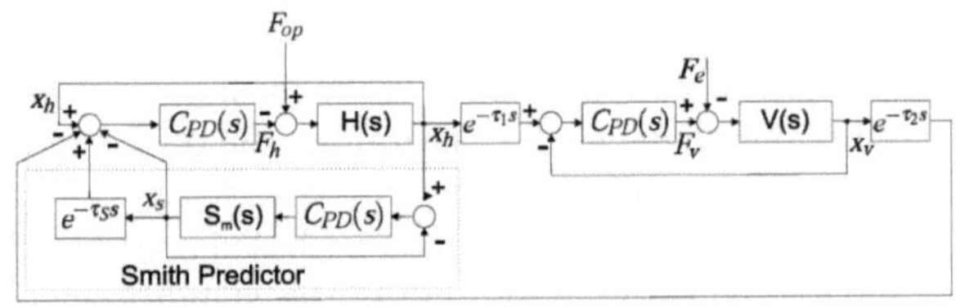

Fig. 15.3 Haptic control scheme including Smith predictor

$$F_h(t) = \underbrace{K_d(\dot{x}_h(t) - \dot{x}_v(t-\tau_2))}_{\text{delayed D-action}} + \underbrace{K_p(x_h(t) - x_v(t-\tau_2))}_{\text{delayed P-action}}, \tag{15.3}$$

$$F_v(t) = \underbrace{K_d(\dot{x}_v(t) - \dot{x}_h(t-\tau_1))}_{\text{delayed D-action}} + \underbrace{K_p(x_v(t) - x_h(t-\tau_1))}_{\text{delayed P-action}}, \tag{15.4}$$

where τ_1, τ_2 are the forward and backward finite delays and K_p, K_d are the PD control gains.

Figure 15.2 presents the control scheme of the haptic system. The robot axes are considered to be decoupled, i.e. the matrices from (15.1)–(15.2) are considered to be diagonal.

Smith Predictor and Stability Insights for the Constant Delay Case

Figure 15.3 presents the control scheme for the considered haptic system. For the stability analysis in presence of time-delays (τ_1, τ_2) the method proposed by [14, 16] will be applied. Using a geometrical approach, this method allows drawing the stability regions in the parameter-space defined by the controller gains. For a self-contained presentation, a summary of the method is presented in the sequel. More precisely, the stability of the closed-loop system is given by the locations of the zeros of the characteristic function $\Delta : \mathbb{C} \times \mathbb{R} \times \mathbb{R} \times \mathbb{R}_+ \to \mathbb{C}$ given by:

$$\begin{aligned}\Delta(s; K_p, K_d, \tau) = {} & (1 + P(s)(1 + e^{-s\tau})(K_p + K_d s)) \\ & (1 + P(s)(1 - e^{-s\tau})(K_p + K_d s)),\end{aligned} \tag{15.5}$$

which has an infinite (but, countable) number of roots (see, for instance [17]). It is worth mentioning that the roots of such a characteristic function are continuously dependent of the system's parameters, and a loss or gain of stability can be detected if characteristic roots cross the imaginary axis (see [6] and the references therein).

Next, for a given $\tau \in \mathbb{R}_+$, and for a frequency $\omega \in \mathbb{R}_+ \setminus \{k\pi/\tau; k \in \mathbb{N}_+\}$ the corresponding crossing points are given by:

$$\begin{cases} K_p^{\pm} = -\Re\Big(1/\left(P(j\omega)(1 \pm e^{-\tau j\omega})\right)\Big) \\ K_d^{\pm} = -\frac{1}{\omega}\Im\Big(1/\left(P(j\omega)(1 \pm e^{-\tau j\omega})\right)\Big) \end{cases}, \tag{15.6}$$

otherwise:

$$\begin{cases} (K_p,\ K_d) = (K_p^{+},\ K_d^{+}), & \text{for } \omega = 2k\pi/\tau, \\ (K_p,\ K_d) = (K_p^{-},\ K_d^{-}), & \text{for } \omega = (2k+1)\pi/\tau, \end{cases} \tag{15.7}$$

where (K_p^+, K_d^+) represents the solution set corresponding to the first part of (15.5), and (K_p^-, K_d^-) correspond to the second part.

Then, when ω varies within some interval $\Omega_l \subset (0, \infty)$, (15.6)–(15.7) define appropriate curves that separate the gain-parameter space in several regions such that each region has the same number of unstable characteristic roots (see, for instance [6], for a more detailed discussion on the so-called D-decomposition method). For a different method, the reader is referred to [18].

15.3 Distributed Delays

In this case, the use of a Smith predictor projects the problem in some appropriate non-delayed case. For $\tau_S = \tau_1 + \tau_2$, $P(s,\ K_p,\ K_d,\ \tau_1,\ \tau_2,\ \tau_S)$ becomes $P(s,\ K_p,\ K_d) = P_1(s,\ K_p,\ K_d)$. According the theoretical results as well as from the practical experiments, in this case the values of K_p and K_d are basically *free* to be chosen from the positive domain ($\mathbb{R}_+$).

In the next paragraphs, the most common delays distributions will be considered and appropriately analyzed in the case of the proposed haptics system. When the time-delays are represented as distributions, Eqs. (15.3)–(15.4) become:

$$\begin{aligned} F_h(t) = {} & \underbrace{K_d(\int_R f(\theta)(\dot{x}_h(t) - \dot{x}_v(t - \tau_0 - \theta))d\theta}_{\text{delayed D-action}} \\ & \underbrace{+K_p \int_R f(\theta)(x_h(t) - x_v(t - \tau_0 - \theta))d\theta,}_{\text{delayed P-action}} \end{aligned} \tag{15.8}$$

$$\begin{aligned} F_v(t) = {} & \underbrace{K_d \int_R f(\theta)(\dot{x}_v(t) - \dot{x}_h(t - \tau_0 - \theta))d\theta}_{\text{delayed D-action}} \\ & \underbrace{+K_p \int_R f(\theta)(x_v(t) - x_h(t - \tau_0 - \theta))d\theta,}_{\text{delayed P-action}} \end{aligned} \tag{15.9}$$

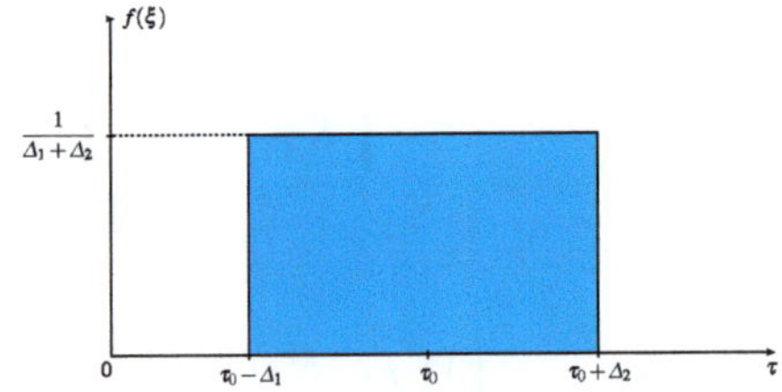

Fig. 15.4 Uniform distribution

where f represents the distribution kernel. If F denotes the Laplace transformation of the distribution kernel, Eqs. (15.10)–(15.11), describing the system response, become:

$$X_h(s) = H(s)\Big(F_{op}(s) - C_{PD}(s)\big(X_h(s) - F(s,\ \tau_2)X_v(s) + (e^{-\tau_S s} - 1)X_S(s)\big)\Big), \tag{15.10}$$

$$X_v(s) = V(s)\left(-F_e(s) + C_{PD}(s)\left(-X_v(s) + F(s,\ \tau_1)X_h(s)\right)\right). \tag{15.11}$$

With the considerations above, the general form of $P(s,\ K_p,\ K_d,\ \tau_1,\ \tau_2,\ \tau_S)$, rewrites as follows:

$$P(s,\ K_p,\ K_d,\ \tau_1,\ \tau_2,\ \tau_S) = P_1(s, K_p, K_d)\left(F(s,\ \tau_2)F(s,\ \tau_1) - e^{-\tau_S s}\right) + P_2(s, K_p, K_d) \tag{15.12}$$

Uniform Distribution. In this case, the uncertainty will be considered to have a random variation, with no additional information about. In other words, the delay will be assumed to have some minimal and maximum values between which it can oscillate. Figure 15.4 presents the distribution of an uniform delay.

Considering $\Delta_1, \Delta_2 > 0$ and $\tau_0 \geq \Delta_1$, the uniform distribution kernel is given by:

$$f(\xi) = \begin{cases} \frac{1}{\Delta_1+\Delta_2}, & \text{if } \tau_0 - \Delta_1 < \xi < \tau_0 + \Delta_2, \\ 0, & \text{otherwise,} \end{cases} \tag{15.13}$$

where τ_0 is the reference delay and Δ_1, Δ_2 represent the delay uncertainty in both directions.

The Laplace transform of the uniform distribution is given by: $F(s) = (e^{-s(\tau_0-\Delta_1)} - e^{-s(\tau_0+\Delta_2)})/(s(\Delta_1+\Delta_2))$. In the proposed Smith predictor scheme, the delay will be considered as $\tau_s = 2\tau_0$ since it must compensate the effects induced by the round trip delay $\tau_1 + \tau_2$, where $\tau_1, \tau_2 \in (\tau_0 - \Delta_1,\ \tau_0 + \Delta_2)$. With these considerations, Eqs. (15.10)–(15.11) rewrite as follows:

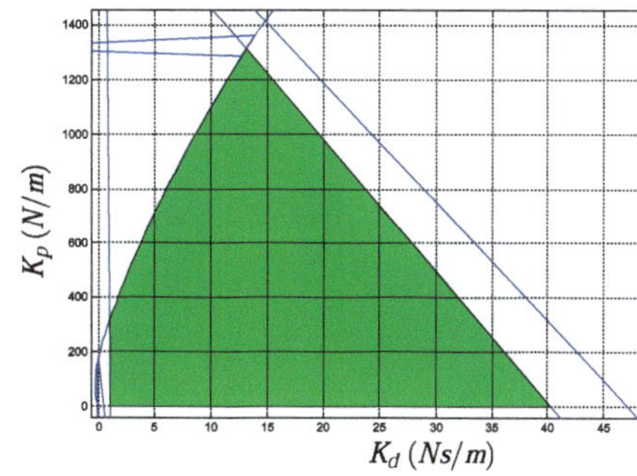

Fig. 15.5 Stability region (in *green*) for (K_p, K_d) for $\Delta_1 = \Delta_2 = 0.015\,\mathrm{s}$

$$
\begin{aligned}
X_h(s) = & \ H(s)F_{op}(s) \\
& - H(s)C_{PD}(s)\left(X_h(s) - \frac{e^{-s(\tau_0-\Delta_1)} - e^{-s(\tau_0+\Delta_2)}}{s(\Delta_1+\Delta_2)}X_v(s) + (e^{-\tau_S s}-1)X_S(s)\right), \\
X_v(s) = & \ V(s)\left(-F_e(s) + C_{PD}(s)\left(-X_v(s) + \frac{e^{-s(\tau_0-\Delta_1)} - e^{-s(\tau_0+\Delta_2)}}{s(\Delta_1+\Delta_2)}X_h(s)\right)\right)
\end{aligned}
\quad (15.14)
$$

and Eq. (15.12) becomes:

$$
\begin{aligned}
P(s,\ K_p,\ K_d,\ \tau_1,\ \tau_2,\ \tau_S) = & \ P_1(s, K_p, K_d)\left(\left(\frac{e^{-s(\tau_0-\Delta_1)} - e^{-s(\tau_0+\Delta_2)})}{s(\Delta_1+\Delta_2)}\right)^2 - e^{-\tau_S s}\right) \\
& + P_2(s, K_p, K_d).
\end{aligned}
\quad (15.15)
$$

Using the method presented in Sect. 15.2, the stability zone in $(K_p,\ K_d)$ parameters is drawn in Fig. 15.5 for the delay uncertainties $\Delta_1 = \Delta_2 = 0.015\ sec$.

Gamma Distribution with Gap. Studies on the problem of controlling objects over communication networks [19] pointed out that, in this case, the time-delays can be modeled by a gamma distribution with gap (see [20] for further details). In such a case, the kernel f is given by:

$$
f(\xi) = \begin{cases} 0, & \xi < \tau, \\ \frac{(\xi-\tau)^{n-1}\mathrm{e}^{-\frac{\xi-\tau}{T}}}{T^n(n-1)!}, & \xi \geq \tau, \end{cases} \quad (15.16)
$$

where $n \in \mathbb{N},\ T > 0$ and $\tau \geq 0$. Note that $f(\xi) \geq 0$ for all $\xi \geq 0$ and $\int_0^\infty f(\xi)\,d\xi = 1$. The gap is defined by τ, and the corresponding *average delay* of (15.16) satisfies: $\tau_m = \int_0^\infty \xi f(\xi)\,d\xi = \tau + nT$. In haptics, the delays' averages are given by: $\tau_1 = \tau_{1_m} = \hat{\tau}_1 + n_1 T_1$ and $\tau_2 = \tau_{2_m} = \hat{\tau}_2 + n_2 T_2$. Figure 15.6 presents an example of delay shape for a gamma distribution with gap. As $T \to 0+$, the kernel (15.16) tends to a Dirac impulse centered at $\xi = \tau$ and one recovers the original system with a point-wise delay τ. The transition to $T = 0$ is smooth from a stability point of

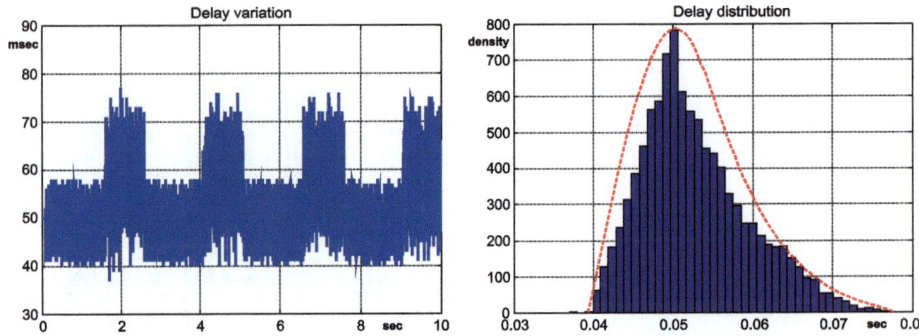

Fig. 15.6 Example of gamma with gap distributed delay: shape and distribution ($\tau = 0.04$ s, $T = 0.01$ s, $n = 1$)

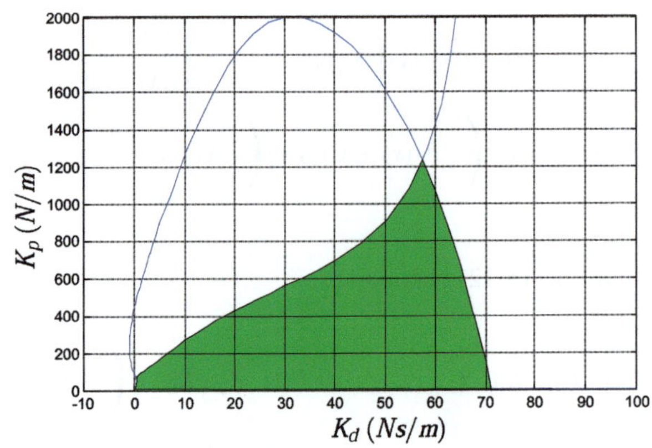

Fig. 15.7 Stability region (in *green*) for (K_p, K_d) for a gamma distributed delay with gap ($\tau = 0.04$ s, $T = 0.01$ s, $n = 1$)

view, as the eigenvalues determining stability are continuous w.r.t. $T \geq 0$. Laplace transform of the gamma distribution with gap is given by: $L(f) = e^{-s\tau}/(1 + sT)^n$. With these considerations, in the case of gamma distribution with gap, Eqs. (15.10)–(15.11) rewrite as follows:

$$X_h(s) = H(s)\left(F_{op}(s) - C_{PD}(s)\left((X_h(s) - \frac{e^{-\tau_2 s}}{(1+sT_2)^{n_2}}X_v(s)) + (e^{-\tau_S s} - 1)X_S(s))\right)\right)$$

$$X_v(s) = V(s)\left(-F_e(s) + C_{PD}(s)\left((-X_v(s) + \frac{e^{-\tau_1 s}}{(1+sT_1)^{n_2}}X_h(s))\right)\right).$$

Consider now $\tau_s = \tau_1 + \tau_2$, with $\tau_1 = \tau_2 = \tau + nT$ ($\hat{\tau_1} = \hat{\tau_2} = \tau$, $n_1 = n_2 = n$, $T_1 = T_2 = T$). Then, Eq. (15.12) rewrites as follows: $P(s, K_p, K_d, \tau, T, n, \tau_S) = P_1(s, K_p, K_d)\left(\left(e^{-\tau s}/(1+sT)^n\right)^2 - e^{-\tau_S s}\right) + P_2(s, K_p, K_d)$. Using the same stability method proposed in Sect. 15.2, Fig. 15.7 presents the stability zone in K_p and K_d parameters space, considering $\tau = 0.04$ s, $T = 0.01$ s and $n = 1$.

Normal Distribution. For this case, similar to the previous ones, a fixed delay is considered for the Smith predictor and a variable delay on the system's side. The system's delays have the following form: $\tau_1 = \tau_2 = \tau_0 + \Delta$, where τ_0 is the nominal delay. The variation of Δ is assumed to be completely random, varying inside some fixed and known interval (similar to a white noise). In the proposed Smith predictor scheme, the delay will be considered $\tau_s = 2\tau_0$ since the effect of the round trip delay $\tau_1 + \tau_2$ has to be compensated. The uncertainty Δ is considered to have a stochastic variation defined by some probability kernel f given by:

$$f(x) = \frac{1}{\sqrt{2\pi\delta^2}} e^{-\frac{x^2}{2\delta^2}}, \tag{15.17}$$

where δ^2 defines the standard deviation. The Laplace transform of the normal distribution (see [21] for further details) is given by:

$$F(s) = \int_0^{\infty} e^{-st} s^{-t^2} dt = \frac{\sqrt{\pi}}{2} e^{\frac{s^2}{4}} erf\left(\frac{s}{2}\right), \tag{15.18}$$

where $erf(\cdot)$ is the complementary error function, defined as:

$$erf(x) = \frac{2}{\sqrt{\pi}} \int_0^x e^{-u^2} du = \frac{2}{\sqrt{\pi}} \left(x - \frac{x^3}{3.1!} + \frac{x^5}{5.2!} - \frac{x^7}{7.3!} + \cdots \right). \tag{15.19}$$

Since the Laplace transform of the normal distribution has a *complex* form, in order to avoid elaborate calculations, a mean stability analysis is proposed. Since the mean value of Δ is zero, the analysis made in the known delay case still holds but for the mean values of $x_h(t)$ and $x_v(t)$ [22]. Therefore, the choice (K_p, K_d) guaranteeing the stability in the known delay case will ensure the mean-stability in the uncertain delay case. In other words, the Smith predictor is able to compensate the delay effects on any given time interval even if for point-wise values of Δ the system is unstable.

Using a Smith predictor for compensating the virtual object without taking into account the other objects in the virtual scene will solve the problem in free motion but for the restricted motion case will induce a poor impact sensation and even an unstable behavior in same cases. This fact is due to the model inconsistency when passing from free into restricted motion. In [23] a switching method between the two models (corresponding to the two situations) is proposed in the teleoperation framework. The main idea is to change the predictor's model based on the tracking error between the haptic and the virtual position, i.e. to assume that there is contact when the error tracking is increasing over a certain limit. In haptics, in general, there is more information available which can be used to ameliorate the systems performances. For instance, the virtual reality can provide some precise information

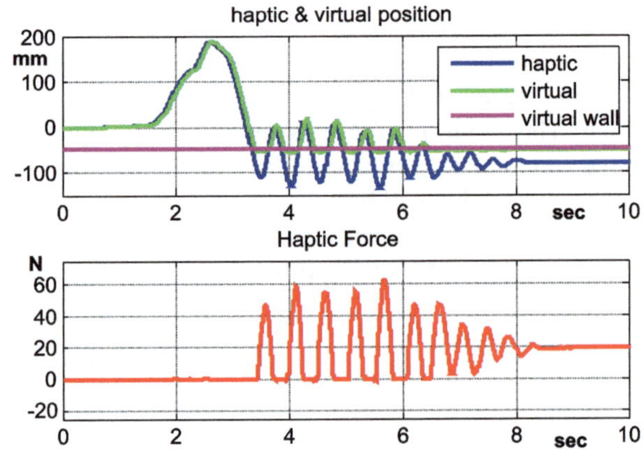

Fig. 15.8 Oscillating behavior of the system when passing from free to restricted motion, using *classic* Smith predictor control

about the neighborhood of the controlled object in order to anticipate the contacts (corresponding to some model switching).

In the case of uncertain delays, it is very important to have an accurate model in the Smith predictor, in order to guarantee the stability of the system. The problem in haptics is that there exist two situations—free and restricted motion, which implies also two different models for the Smith predictor. In the next Section, a specific solution for haptics is proposed in order to improve the Smith predictor.

15.4 Smith Predictor with Distance Feedback

In the previous Section it was shown that by using a Smith predictor good results can be obtained for different types of delays' distributions. These results correspond to the free motion case. In restricted motion the results are no longer valid since some contact force must be added in the dynamics. The problem is linked to the model used in the Smith predictor which is no more similar to the virtual reality model corresponding to the contact case. Unfortunately, the basic form of the Smith predictor can compensate the delays only in free motion and, in restricted motion, the model is no longer *sufficient*, since the virtual environment is not included and will induce a poor impact sensation and, in some cases, even an unstable behavior. Figure 15.8 illustrates the system behavior when passing from free to restricted motion using a *"classical"* Smith predictor.

The *classical* Smith predictor used for this case writes as following:

$$M_s(x_s)\ddot{x}_s(t) + C_s(x_s, \dot{x}_s)\dot{x}_s = -F_s(t), \tag{15.20}$$

where x_s is the predicted object position, F_s is the predicted force control signal, M_s is the symmetric and positive-definite inertia matrix and C_s is the Coriolis matrix, both equal to the virtual reality object mass and Coriolis matrix, respectively. A method to *update* the system model used in the predictor is proposed, using the distance until collision available in the virtual reality scene.

The main idea of the proposed solution is to introduce into the Smith predictor the environmental force F_{es} by using the distance between the controlled virtual object and other objects in the scene. Based on the information received by the virtual environment it is possible to predict the impact moment and to *update* the predictor's model resulting in some more accurate system. The information is also affected by the communication delays but this does not represent a critical problem. Even if some changes appear during the delay-time period, the model will still be able to ensure the desired behavior [24]. In Sect. 15.5, these situations will be experimentally presented. Introducing the environmental force in the predictor, Eq. (15.20) rewrites as follows:

$$M_s(x_s)\ddot{x}_s(t) + C_s(x_s, \dot{x}_s)\dot{x}_s = -F_s(t) + F_{es}, \tag{15.21}$$

where F_{es} represents the estimated environmental force calculated based on the distance extracted from the virtual reality: $F_{es} = K_{wall}(P_s - \widetilde{P}_{wall}) + B_{wall}\dot{P}_s$, where K_{wall} and B_{wall} represent the stiffness and damping used to compute the virtual force environment, $\widetilde{P}_{wall}$ is the virtual wall position received from the virtual scene (x, y, z) and P_s, $\dot{P}_s$ are the virtual object position and velocity estimations used in the Smith predictor. In the next Section, different experiments will be presented in order to validate this approach.

15.5 Experimental Validation

In order to guarantee a full control of the communication delays and processing time, all the control algorithms (for haptic interface/virtual object) and virtual environment simulations will be run on the same computer. The haptic interface, Fig. 15.9a and b, consists of three direct-drive motors and three optical quadrature encoders with 1000 pts/rev (with a gear ratio of 1/10). The controllers and the virtual simulation are running in real time mode (on RTAI Linux) with a sampling time of 1 ms. Figure 15.9c and d illustrates the two virtual scenes (*simple* environment/virtual box) that will be used and the virtual object. The virtual object is modeled to be spherical mass (equal to the haptic interface mass) ($M_h = M_v$). The environmental force (F_{e_x}, F_{e_y}, F_{e_z}) resulting in case of an impact with the virtual environment (virtual walls) is defined by the following equation: $F_e = K_{wall}(P_v - P_{wall}) + B_{wall}\dot{P}_v$, with: $F_e = \left(F_{e_x}\ F_{e_y}\ F_{e_z}\right)^T$, $P_v = (x_v\ y_v\ z_v)^T$, $P_{wall} = (x_{wall}\ y_{wall}\ z_{wall})^T$, where $K_{wall} = 20000$ N/m and $B_{wall} = 10$ Ns/m represent the stiffness and damping used to compute the virtual force environment, P_{wall} is the virtual wall position and P_v, $\dot{P}_v$ are the virtual object position and velocity.

All the experiments are carried out by a human operator explaining why the conditions will not be exactly the same in all experiments. The testing scenarios

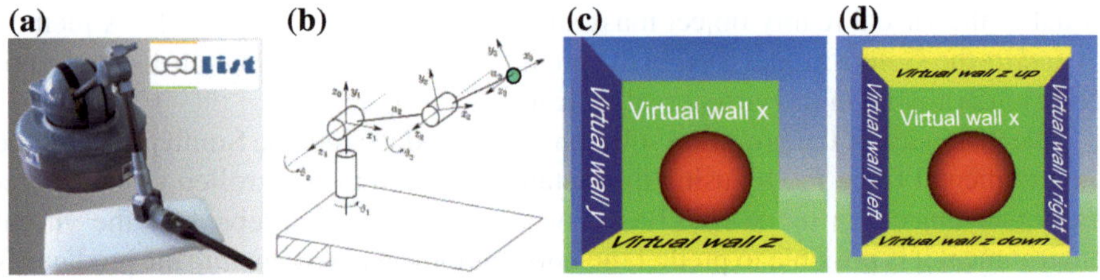

Fig. 15.9 Haptic system. **a** Haptic interface, **b** robot geometry, **c** *simple* virtual scene, **d** virtual box scene

are the same for each experimental category: (i) free motion (random motions on each axis) and restricted motion (wall contact on each axis), (ii) contact with moving objects, i.e. a sinusoidal moving wall will be used in order to provide a more realistic scenario, (iii) free and restricted motion inside a virtual box (random motions with or without contact on each axis and multi-point contacts). The purpose of the last scenario is to show that the proposed method can be adapted for more complex cases than just a simple wall, where the virtual scene has fast changes and complex shapes. More precisely, this experiment aims to demonstrate the efficiency of the proposed method when fast changes are applied i.e. for *small* working spaces where contacts can arrive very often. For such situations the system must be able to act rapidly in order to provide a realistic feeling—high degree of transparency. During this experiment, moving sides are used in order to simulate more complex situations (like tiny environment with variable size/position) and thus to demonstrate the method's efficiency. A complete experimental illustration can be found in [25].

In the case of *fixed and known delays*, the problem is reduced to the free of delay case and considered because the Smith predictor fully compensates the system's delays. A complete experimental illustration of this case can be found in [25]. It is worth mentioning that, as expected, the results are perfectly similar to the *ideal* case. More precisely, in free motion the feedback force is *small*, (force which is not felt by the human operator) and in case of hard contacts (wall contacts) the force increases very fast providing the contact sensation in a very accurate way.

Uniform distributed delays. For the next example, based on the practical needs in terms of transparency and respecting the theoretical results from Fig. 15.5, the following controller gains will be used: $K_p = 1000$ N/m, $K_d = 15$ Ns/m. For this case the delay starts at 35 ms (round trip 70 ms) for 2.5 s and then switches to 65 ms (round trip 130 ms) for another 2.5 s, and so on. A complete experimental illustration can be found in [25]. In free motion the results are closed to the *ideal* case, while in restricted motion, the contact feeling is provided in an accurate way. In this chapter, the moving object case will be shown since it presents the most intersecting situation. More precisely, a sinusoidal moving wall is used in order to illustrate the system's behavior in the case of contacts with virtual moving objects. Figure 15.10 presents the results.

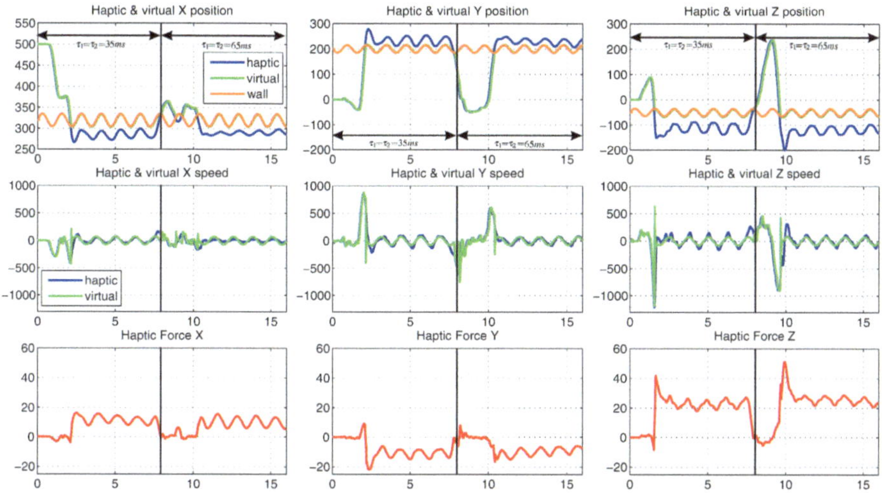

Fig. 15.10 Restricted motion with sinusoidal moving object—uniform distributed time-delays

The goal of this experiment is to show that the results in terms of transparency are similar to the ones obtained in the case of *simple* contacts and moreover, if the contact is maintained, the user is able to feel the moves of the object in an accurate and natural way. The proposed approach from Sect. 15.4 works correctly even if the objects' position is varying.

Gamma distributed delays with gap. The delay's shape used for the next experiments is the one presented in Fig. 15.6. According to Fig. 15.7, the following controller's gains will be used: $K_p = 1100$ N/m, $K_d = 55$ Ns/m. In the case of free and restricted motion, compared with the result obtained for uniform distribution the performances are better (see [15, 25] for complete experimental illustration). The contact feeling is provided in an accurate way and the transition from free to restricted motion is closed to the ideal case. Since the delay variation is more restrictive in this case, the choice for the controller's gains (K_p, K_d) has increased and the allowable values are *closer* to the *ideal* case. Similar to the previous case, the contact with a moving objects will be analyzed. Figure 15.11 depicts the sinusoidal moving wall impact for gamma distribution with gap time-delays.

As expected, the results are better than in the uniform distribution case. The main difference is the tracking error which is smaller, result that is reflected also in a better contact feeling.

Normal distributed delays. According to the theoretical results from Sect. 15.3, the choice for parameters (K_p, K_d) can be made similar to the *ideal* case, i.e. is basically *free* as long as the values are positive. Based on this, the controller's gains are chosen equal to the ones used in the *ideal* case: $K_p = 1800$ N/m, $K_d = 80$ Ns/m. The results for free and restricted motions are close to the *ideal* case for both situations: free and restricted motion—the viscosity effect is low, the impact is stiff and the tacking

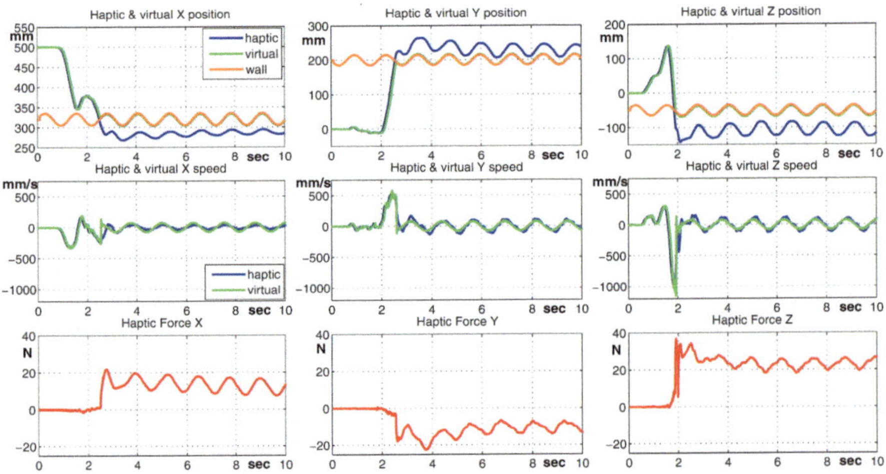

Fig. 15.11 Restricted motion with sinusoidal moving object—gamma with gap distributed time-delays

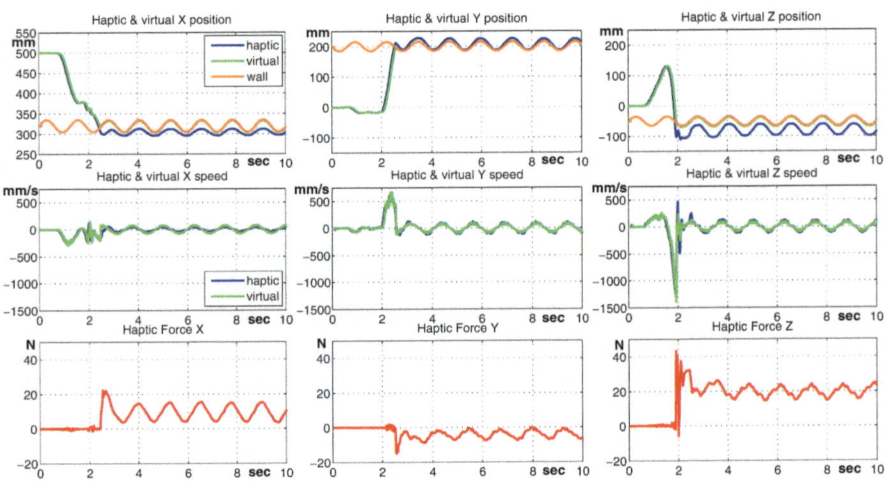

Fig. 15.12 Restricted motion with sinusoidal moving object—normal distributed time-delays

error is small. The case of contacts with moving objects (sinusoidal moving wall) is depicted in Fig. 15.12.

The results are consistent with the reality and, as suspected, *close* to the *ideal* case, since the analysis of this case leads to the non-delayed case.

Uncertainties in the feedback used in the Smith predictor. Here, a special experiment is proposed in order to validate the proposed method under perturbations on the feedback. Some uncertainties on the wall position feedback used in the Smith predictor will be added. More precisely, a random variation will be added in order to

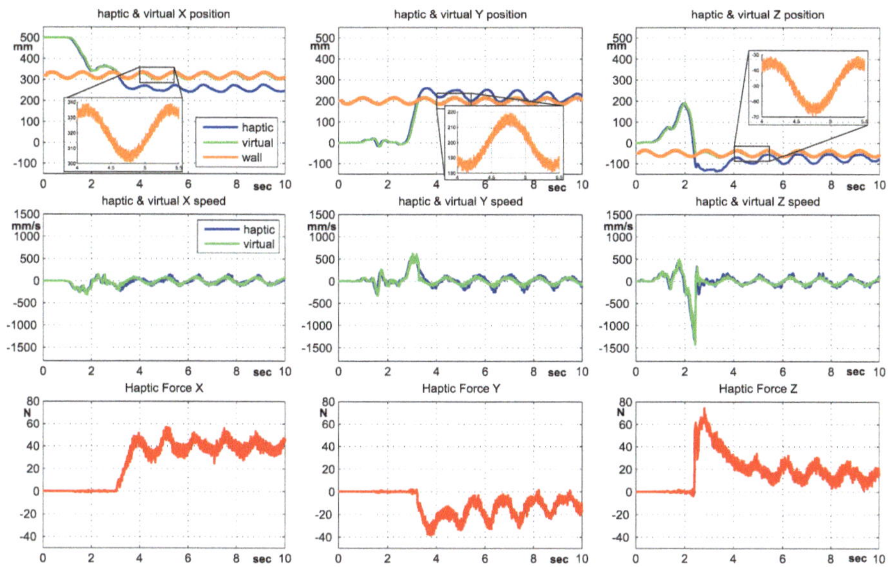

Fig. 15.13 Restricted motion case with sinusoidal moving object, random uncertainties on the Smith predictor side, under normal distributed time-delays

alter the delayed information used in the Smith predictor. In our opinion, this random variation is the worst case, because the uncertainty value has *big* variations (random) and the frequency is high. Figure 15.13 depicts this situation for the sinusoidal *moving wall* and under normal distributed time-delays.

The oscillating force denotes some small vibrations at the haptic end, which can be slightly felt by the human operator, without providing disturbing behaviors. Increasing the uncertainty value interval and maintaining the same frequency, the vibrations can become annoying.

All the experiments have been done by a human operator and this explains the reasons for which the conditions are not perfectly the same in all experiments. The testing scenarios were the same for each experimental category (free motion, restricted motion and virtual box). By using a human operator for carrying out the experiments, an important experience regarding the perception in terms of transparency and feeling was earn.

15.6 Concluding Remarks

In this chapter, a new control method for the haptic delayed systems was presented. The three most common delay distributions (uniform distribution, gamma distribution with gap and normal distribution) were analyzed from the stability point of view and tested on a 3-dof real-time experimental platform. The varying time-delays can be modeled in order to choose more efficiently the controller's parameters providing high performance in terms of end user perception and guarantying the system's stability. As presented in Sect. 15.3, depending on the distribution type, the choice of controller's gains can be more or less restrictive. More precisely, for the uniform distribution, the choice of parameters appears to be the most restrictive one, while for the normal distribution the choice is close to the *ideal* case. The uniform distribution corresponds to the case when little information about the delay variation is known (i.e. only the minimum and maximum values are needed in this case). On the other hand, if the delay shape can be approximated by the gamma with gap or normal distribution the parameters' choice will be more important. Thus, the uniform distribution appears to give the most restrictive stability zone, based on a minimal information and the other two (gamma with gap and normal) distributions give a less restrictive stability zone, but based on more detailed information.

In future works, an extension of this approach will be presented for more complex virtual scenes like concave or non-regular geometric shapes and also for a haptic interface with six degrees of freedom.

Acknowledgments The work of Bogdan Liacu was financially supported by CEA, LIST, Interactive Robotics Laboratory, Fontenay aux Roses, F-92265, France.

References

1. Sreng, J., Lécuyer, A., Mégard, C., Andriot,C.: Using visual cues of contact to improve interactive manipulation of virtual objects in industrial assembly/maintenance simulations. IEEE Trans. Visual Comput. Graphics **12**(5), 1013–1020 (2006)
2. Gosselin, F., Megard, C., Bouchigny, S., Ferlay, F., Taha, F., Delcampe, P., D'Hauthuille, C.: A VR training platform for maxillo facial surgery. Applied Human Factors and Ergonomics (AHFE) International Conference, Miami, Florida, USA, Advances in Cognitive Ergonomics (2010)
3. Andriot, C., Perret, J.: Immersive virtual prototyping with haptic feedback and virtual manikins. In: Proceedings of IDMME - Virtual Concept, Beijing, China (2008)
4. Hokayem, P.F., Spong, M.W.: Bilateral teleoperation: an historical survey. Automatica **42**(12), 2035–2057 (2006)
5. Loiseau, J.-J., Michiels, W., Niculescu, S.-I.: Topics in time delay systems: analysis, algorithms and control. Lecture Notes in Control and Information Sciences. Springer, Berlin (2009)
6. Michiels, W., Niculescu, S.-I.: Stability and stabilization of time-delay systems: an Eigenvalue-based approach. Advances in design and control, Society for Industrial and Applied Mathematics (2007)
7. Keller, A.A.: Time-Delay Systems: With Applications to Economic Dynamics and Control. LAP LAMBERT Academic Publishing, Saarbrucken (2011)
8. Cushing, J., Cushing, J.: Integrodifferential equations and delay models in population dynamics. Lecture Notes in Biomathematics. Springer, Berlin (1977)
9. Sipahi, R., Atay, F.M.: Stability of traffic flow behavior with distributed delays modeling the memory effects of the drivers. SIAM J. Appl. Math. **68**(3), 738–759 (2007)
10. MacDonald, N.: Biological Delay Systems: Linear Stability Theory. Cambridge University Press, Cambridge (1989)
11. Yang, S.: Internet-Based Control Systems: Design and Applications. Springer, Berlin (2011)
12. Smith, O.J.M.: Closer control of loops with dead time. Chem. Eng. Prog. **53**(5), 217–219 (1957)

13. Morărescu, I.-C., Niculescu, S.-I., Gu, K.: On the geometry of stability regions of smith predictors subject to delay uncertainty. IMA J. Math. Control Inf. **24**(3), 411–423 (2007)
14. Liacu, B., Mendez-Barrios, C., Niculescu, S.-I., Olaru, S.: Some remarks on the fragility of pd controllers for siso systems with i/o delays. In: 14th International Conference on System Theory and Control, Sinaia, Romania (2010)
15. Liacu, B., Morărescu, I.-C., Andriot, C., Niculescu, S.-I., Dumur, D., Boucher P., Colledani F.: Some remarks on smith predictor-based control with distance feedback for a 3-dof haptic system with distributed delays. In: Proceedings of IFAC Joint Conference (5th Symposium on System Structure and Control, 11th Workshop on Time-Delay Systems, 6th Workshop on Fractional Differentiation and Its Applications), Grenoble (2013)
16. Mendez-Barrios, C., Niculescu, S.-I., Morărescu, I.C., Gu, K.: On the fragility of pi controllers for time-delay siso systems. In: 16th Mediterranean Conference on Control and Automation, pp. 529–534 (2008)
17. Niculescu, S.-I.: Delay effects on stability: a robust control approach. Springer, Heidelberg, LNCIS (2001)
18. Liacu, B., Taha Koru, A., Özbay, H., Niculescu, S.-I.: Optimizing low-order controllers for haptic systems under delayed feedback. Control Eng. Pract. **21**, 655–668 (2013)
19. Roesch, O., Roth, H., Niculescu, S.-I.: Remote control of mechatronic systems over communication networks. In: IEEE International Conference on Mechatronics and Automation, vol. 3, pp. 1648–1653 (2005)
20. Morărescu, I.-C., Niculescu, S.-I., Gu, K.: Stability crossing curves of shifted gamma-distributed delay systems. SIAM J. Appl. Dyn. Syst. **6**(2), 475–493 (2007)
21. Whittaker, E.T., Watson, G.N.: A Course in Modern Analysis, 4th ed. Cambridge University Press, Cambridge (1990)
22. Herzog, F.: Lecture on Stochastic Systems. ETH Zurich, Zurich (2010)
23. Ching, H., Book, W.: Internet-based bilateral teleoperation based on wave variable with adaptive predictor and direct drift control. J. Dyn. Syst. Meas. Control **128**, 86–93 (2006)
24. Liacu, B., Morărescu, I.-C., Andriot, C., Niculescu, S.-I., Dumur, D., Boucher, P., Colledani, F.: Some remarks on the fragility of smith predictors used in haptics. In: 11th International Conference on Control, Automation and Systems (ICCAS), pp. 1851–1856 (2011)
25. Liacu, B.: Network-based Haptic Systems with Time-Delays. Ph.D. dissertation, Supélec (2012)

[illegible]

18. [illegible]

19. [illegible]

[illegible]

[illegible]

22. [illegible]

23. [illegible]

[illegible]

[illegible]

[illegible]

[illegible]

Index

A. Seuret et al. (eds.), *Low-Complexity Controllers for Time-Delay Systems*,
Advances in Delays and Dynamics 2, DOI: 10.1007/978-3-319-05576-3,

MIX
Papier aus verantwortungsvollen Quellen
Paper from responsible sources
FSC® C105338

If you have any concerns about our products,
you can contact us on
ProductSafety@springernature.com

In case Publisher is established outside the EU,
the EU authorized representative is:
Springer Nature Customer Service Center GmbH
Europaplatz 3, 69115 Heidelberg, Germany

Printed by Libri Plureos GmbH
in Hamburg, Germany